Conquering
Math Anxiety

A SELF-HELP WORKBOOK

Cynthia A. Arem, Ph.D.
Pima Community College

THIRD EDITION

BROOKS/COLE
CENGAGE Learning™

Australia • Brazil • Japan • Korea • Mexico • Singapore • Spain • United Kingdom • United States

BROOKS/COLE
CENGAGE Learning™

**Conquering Math Anxiety:
A Self-Help Workbook, Third
Edition**

Cynthia A. Arem

Publisher: Charlie Van Wagner

Developmental Editor:
Natasha Coats

Assistant Editor:
Stefanie Beeck

Editorial Assistant:
Mary de la Cruz

Media Editor: Sam Subity

Marketing Manager:
Greta Kleinert

Marketing Assistant:
Angela Kim

Marketing Communications
Manager: Katherine Malatesta

Content Project Manager:
Michelle Cole

Creative Director: Rob Hugel

Art Director: Vernon Boes

Print Buyer: Linda Hsu

Rights Acquistions Accounts
Manager, Text: Bob Kauser

Rights Acquistions Accounts
Manager, Image:
Amanda Groszko

Production Service: Scratchgravel Publishing Services

Copy Editor: Frank Hubert

Illustrator: Scratchgravel
Publishing Services

Cover Designer: Ross Carron

Cover Image: IPNstock

Compositor: Scratchgravel
Publishing Services

For product information and technology assistance, contact us at
Cengage Learning Customer & Sales Support, 1-800-354-9706

For permission to use material from this text or product, submit
all requests online at **www.cengage.com/permissions**
Further permissions questions can be emailed to
permissionrequest@cengage.com

Library of Congress Control Number: 2009920017

ISBN-13: 978-0-495-82940-9
ISBN-10: 0-495-82940-4

Brooks/Cole
10 Davis Drive
Belmont, CA 94002-3098
USA

Cengage Learning is a leading provider of customized learning
solutions with office locations around the globe, including Singapore,
the United Kingdom, Australia, Mexico, Brazil, and Japan. Locate your
local office at: **www.cengage.com/global**

Cengage Learning products are represented in Canada by Nelson
Education, Ltd.

To learn more about Brooks/Cole, visit
www.cengage.com/brookscole

Purchase any of our products at your local college store or at our
preferred online store **www.ichapters.com**

Printed in Canada
1 2 3 4 5 6 7 13 12 11 10 09

To Arnie

My best friend, confidant, teacher, guardian angel, and nurturer of all my dreams and ambitions—so loving, sensitive, and sincere. Thank you for being there, every step of the way in my life and in my work.

CONTENTS

6 Win with "Success in Math Visualizations" 74

7 Enhance Your Learning Style 86

8 Effective Math Study Skills: The Fuel of Excellence   109

9 Conquer Test Anxiety   145

Millions of students in this country are terrified of math. They do whatever they can to avoid numbers and math problems as if they were the plague. In school, they put off taking math as long as possible. And when given the choice, they select majors requiring little or no math. If they do have to take the subject, they dread entering the classroom, let alone taking an exam. One young student described how she panicked in math class to the point that she would run out of the room and vomit uncontrollably. As a child, she often had nightmares about numbers chasing her, wanting to hurt her.

High levels of anxiety can devastate a student's ability to perform, resulting in poor academic progress and high dropout rates. Research among college students has shown math generates anxiety reactions among students who are not necessarily highly anxious in other situations. One study at a large midwest university disclosed one-third of all students who requested behavioral counseling complained of anxiety related to math.

As a college counselor, I've seen hundreds of students each semester who fear math or do their best to avoid taking it. In fact, I believe there are people in all walks of life who are math anxious. Invariably, when I'm at social functions or at meetings, people confide in me about their math woes. They often tell me math has always been their biggest problem in school. They say things such as: "If it weren't for math, I'd be going to school right now." "I'd be making a career change." "I'd be more successful than I am right now." "I'd go back to school and get that degree I always wanted." And on and on.

Widely reported studies have shown math serves as a "critical filter" in determining many educational, vocational, and professional options (National Research Council, 1998; Sells, 1978). Math avoiders are finding themselves shut out of today's most rewarding and profitable careers.

College algebra is a minimum math requirement for a bachelor's degree at almost all undergraduate institutions. At the community college level, we see many students who can't possibly meet this math requirement, either because they had poor math backgrounds to begin with or because they've been out of school so long they've forgotten what they once knew. Often, these students must take four math courses just to meet this university requirement for their major. They need to take fundamentals of math—a course reviewing percentages, decimals, and fractions—before they can go on to take beginning algebra, intermediate algebra, and finally, college algebra. For students who are afraid of math, this is a living nightmare. Furthermore, the news media increasingly point out that the majority of Ph.D. degrees in science and engineering given in this country are being awarded to foreign-born graduates who, presumably, are more comfortable with math than U.S. graduates.

Why is math such a problem for Americans? How can we fully understand this problem? What can be done to overcome the devastating effects of math anxiety and math avoidance? I've designed this workbook to answer these questions.

Conquering Math Anxiety: A Self-Help Workbook presents a comprehensive, multifaceted treatment approach to reduce math anxiety and math avoidance. This unique, carefully outlined methodology involves anxiety management and reduction, reduction of internal psychological stumbling blocks, attitude changing, confidence building, "Success in Math Visualizations," learning-style enhancements, problem-solving strategies, and effective study and test-taking skills. In addition, it teaches winning strategies for overcoming test anxiety on math exams. This workbook provides detailed explanations along with a host of varied exercises, methods, and worksheets for helping math-anxious students deal with and overcome math fears.

The CD accompanying this workbook contains recordings of my favorite and most powerful relaxation and visualization exercises for reducing math anxiety. I've used these exercises successfully for many years with math-anxious students, both in my classes and in counseling.

Although *Conquering Math Anxiety* is based on my experience at the community college level, my primary purpose in writing it is to reach *all* math-anxious individuals. Many are attending universities, community colleges, or high schools. Many are no longer in school but find that math fears continue to interfere with the pursuit of their personal goals, often creating lowered self-esteem, frustration, or dissatisfaction.

Because this workbook is comprehensive in its scope, it is relevant for people at any educational level or in any career. It is designed so the reader can simply jump in and draw from it as little or as much as he or she needs.

No prerequisites are needed for completing the activities in this workbook. Mathematics anxiety can be found at any math level, from the fundamentals of math through statistics and beyond. I've even had students taking calculus simultaneously enrolled in my math anxiety reduction course.

Conquering Math Anxiety uses the analogy of a road map detailing the route to math success, but its individual chapters are like recipes that can be extracted and used entirely on their own according to need. For this reason, the workbook and CD are useful by themselves or as supplements to any mathematics course in which a student or teacher identifies a problem area impeding smooth progress.

My background as a college counselor, teacher, and workshop trainer leads me to speak directly to those with whom I work and to involve them thoroughly in the learning process. Therefore, I have included a maximum number of exercises. Also, throughout the book, I address the reader in the first person. After first analyzing the roots of math anxiety, I detail specific techniques for managing anxiety and overcoming psychological barriers to math. This leads to the critical importance of reversing negative "math self-talk," creating positive attitudes, and building self-confidence. The section on math success visualization is unique. It helps readers reprogram their thinking to promote math success. Then I focus on learning styles, study skills, and problem-solving approaches as they apply specifically to math. The workbook ends with an in-depth look into techniques for conquering math test anxiety and an analysis of the importance of mathematics in all aspects of life. Interspersed throughout the workbook you will find humorous cartoons with a positive slant.

What's New in the Third Edition

The most important changes to this edition are summarized here:

Chapter 2

- New research data have been added to the sections "Poor Teaching Methods" and "Gender Stereotyping and Socialization."
- The section "Dyscalculia: A Cause or a Symptom?" is new. It discusses dyscalculia and research related to it.
- Figure 2-1 "The Math Anxiety Cycle" is expanded and more comprehensive.

Chapter 3

- This chapter is renamed *Learn to Manage Anxiety and Improve Working Memory*.
- The section "Anxiety Can Affect Your Long-Term and Working Memory Systems" is new. This section discusses the roles these memory systems have when students learn math and solve problems. It highlights the important research on the devastating effects anxiety has on working memory and math problem solving.
- Figure 3-1 "Working Memory's Simultaneous Processes Used in Math" is new.

Chapter 4

- The section "Math Self-Concept and Math Self-Efficacy" is new. This section discusses the important roles math self-concept and math self-efficacy have in learning math. Related research data are included.
- Exercise 4-4 "Your Math Self-Concept: A Self-Assessment " is new.

Chapter 7

- The section "A Sequential versus a Global Learning Style" is new. This section discusses the difference between sequential and global learners. It also provides studying suggestions for both learning styles.
- The section "Personal Thinking Styles" is now renamed "A Deductive versus an Inductive Learning Style," and it has been moved earlier in the chapter.

- Exercise 7-2 "Collaborative Learning: Share Your Insights on Learning Styles" is new.

Chapter 8

- The section "From Acquisition to Understanding" is new. This section discusses the four stages of learning: acquisition, fluency, adaptation, and generalization. It encourages students to go beyond the mere acquisition of a new math concept to the later learning stages to increase understanding and math proficiency.
- Figure 8-1 "Math Learning Stages and Strategies" is new. This figure provides suggestions on how to successfully accomplish each learning stage and move to the next higher one.
- The section "Clear the Minefield" is new. It discusses how students' faulty beliefs about studying are really booby traps that prevent them from moving toward their math goal.
- Figure 8-2 "Disarm Your Minefield' is new. It summarizes some common faulty beliefs and offers study strategies for neutralizing them.
- The section "Useful Strategies for Dealing with Dyscalculia" is new. Its title tells it all.
- Figure 8-3 "Suggestions for Dyscalculia Difficulties" is also new. This figure includes suggested strategies to help nine different dyscalculia difficulties.

Chapter 9

- Section "Be Your Own Exam Coach" is new. This section encourages students to be their own coach, give themselves pep talks, keep themselves focused on their task, push away distracting thoughts, and set realistic, positive exam goals.
- Exercise 9-12 "Collaborative Learning: The Best Coaching Practices" is new.
- The section "After Your Next Exam" is new. In this section, students are asked to reflect on what strategies worked best for them on their next exam and what they learned about themselves and their test-taking skills.
- Exercise 9-14 "Characteristics of Optimal Test Performance: A Self-Assessment" is new.

Chapter 10

- The section "Build Metacognitive Strategies" is new. This section asks students to reflect on their own "cognitive" or thinking processes so they can achieve an in-depth understanding of how they learn math and solve problems.
- Exercise 10-3 "Metacognitive Strategies for Success in Math" is new. This exercise provides relevant questions for students to ask themselves at each of the three main stages of metacognition: the planning stage, the monitoring stage, and the reflection/evaluation stage.

Note

Although this book is based on my experience as a community college teacher and counselor, both college and high school teachers are successfully using it in their classes, tutoring, and learning resource centers.

Acknowledgments

I am extremely grateful to those who have assisted me in the development of the ideas presented in this workbook: the many math-anxious students I have counseled and taught, my colleagues who shared ideas and teaching suggestions, and my family and friends who offered support and assistance.

This work would not have been possible if not for the many wonderful students I have worked with over the years—both in counseling sessions and in the math anxiety reduction courses and workshops I have taught. In the process of helping them deal with math anxiety, I have learned much and now feel privileged to be able to share this knowledge with others. My students have been my teachers, and I would like to thank each and every one of them.

The most important person in the development and writing of this book has been my husband, Arnold Arem, M.D. His many hours of dedication to this project have been invaluable to me. To each of the ideas presented here he listened carefully and gave his thoughtful critique. He read and reread my manuscript at each stage of its development and offered me counsel, detailed editing advice, and much needed emotional support. I am particularly delighted that he developed the ideas for the lighthearted cartoons that appear in this work.

I would also like to thank my daughter, Kimba, for the beautiful, relaxing music she composed for the CD accompanying this workbook. Her caring input has been an inspiration. I also extend my sincere appreciation to my son, Keith, and to PCB Productions for their excellent production of the accompanying CD. Keith's dedication to quality and his hard work have helped produce a superb recording.

A dear friend and colleague I want to thank is Debbie Yoklic. Our contacts over the years and experience team-teaching math anxiety reduction courses were the stimulus for writing this book. I am grateful to both Dr. Arlene Scadron for her excellent editorial advice and my colleagues in the Mathematics Department of Pima Community College, West Campus, for their thoughtful suggestions. The late Dr. Frank Rizzuto was an inspiration for many ideas in this workbook.

My special appreciation to Louis Smith for his stimulating discussion with me on one of the great loves in his life: math. I would like to especially thank him for his creative input and practical suggestions.

I would like to thank reviewers Debra Bryant from Tennessee Technological University, Marie Franzosa from Oregon State University, and Roger Werbylo from Pima Community College for their excellent editorial comments. The staff at Cengage Learning also deserve special credit. In particular, I would like to thank Charles Van Wagner, Publisher; Natasha Coats, Developmental Editor; Stephanie Beeck, Assistant Editor; Michelle Cole, Content Project Manager; Greta Kleinert, Senior Marketing Manager; and Angela Kim, Marketing Coordinator. Lastly, I would like to thank Anne Draus and Scratchgravel Publishing Services for their excellent composition and production services.

Cynthia A. Arem, Ph.D.

The Road Map to Success

Have you ever found yourself lost, in a strange or unfamiliar place, without a map to guide you? Did you suddenly feel alone? Did you feel scared, not knowing what to do or which way to turn? Perhaps you stopped to ask for directions, but they were confusing, misguided, or incomprehensible. If the hour was growing late, you might have felt a sense of menace or dread. Perhaps you began to perspire, your muscles tightened, your pulse quickened, your breathing became more rapid, and all your senses heightened, trying to detect if danger was present.

Now let's imagine for a moment you were miraculously given a detailed map of the area, not only showing where you were but also providing explicit instructions for reaching any destination you chose. Every step along the way was made clear, every turn and every landmark carefully spelled out and described. Your map left no stone unturned, no shadow of a doubt as to what direction to take and how to proceed. Wouldn't such a map have made your journey less stressful, more pleasurable, and helped guarantee success in reaching your goal?

Is dealing with math like being lost in an unfamiliar or strange place? Do you sometimes feel as if you're all alone, no one understands what you're going through, and others' efforts to help are confusing or misleading? When doing math, do you ever feel scared, panicky, or have a sense of dread or impending doom—not knowing which way to turn or where to get the assistance you so desperately need?

Let me assure you that you are not alone. There are many people who have felt the same way at one time or another. Students often describe feelings of devastating failure and utter defeat when it comes to doing math. Mathematical reasoning and numbers are their nemesis and have been for much of their school lives. They shun exposure to math as if it were a frightening, unknown danger. Whenever possible, they purposely avoid any subject requiring math and choose majors requiring little or preferably no math. In math classes or even in math tutoring sessions, their minds turn off and they become lost, comprehending little of what is explained or taught.

If you are one of these students—lost, feeling alone, apprehensive, shunning math—you have come to the right place. I have written this workbook specifically for you. I've designed it to be your personal, detailed map to help you find your way, to aid you in overcoming your fear of math, and to ensure that you achieve success in math. Included are detailed explanations and exercises all along your path to illuminate your journey and to make it easier and more fulfilling.

As you follow this road map, you will find many important branch points to explore and experience. We begin our journey in Chapter 1 by first discovering whether you really have math anxiety, and then we analyze your math success goal. From there, we'll look at where your math fears and anxieties initially began and why they have persisted.

Chapter 2 asks you to look at your math anxiety history and identify the factors that influenced you as well as the myths, stereotypes, or games that might have affected you. By diagraming the math anxiety process, you can see how previous negative math experiences combined with negative self-talk lead to overwhelming anxieties and fears.

In Chapter 3, you learn how anxiety affects your memory and ways to control and manage anxiety so it can work for you and not against you. You learn how high, uncontrolled anxiety levels lead to panic and the inability to perform, whereas moderate, properly managed anxiety marshals optimal performance, good memory, and clear thinking.

Chapter 4 explores how you deal with problems that arise in your life and how your coping methods can aid or hinder your efforts to overcome math anxiety. The Wall Fantasy is designed to increase your awareness of what you may be doing to impede your success in math. We also assess your math self-concept.

Next, in Chapter 5, we look at the importance of attitudes and ways of changing negative attitudes toward math to positive ones. I guide you through exercises to help you develop positive, enhancing self-dialogue related to math and your ability to do math.

Then our journey takes us to the very exciting area of "Success in Math" imagery. Research has shown that positive mental imagery can greatly improve the status of our lives mentally, emotionally, and physically. In Chapter 6, you are given excellent visualization techniques to achieve your math goals. You are guided through a beautiful imagery technique that has already advanced the lives of many hundreds of math students who overcame the same math anxieties that you have been experiencing. You will also find this exercise recorded on the accompanying CD.

Chapter 7 looks at your math learning style. Through the use of carefully chosen questions, you are asked to assess the critical factors influencing your ability to learn math. Whether you are a visual, auditory, or kinesthetic/tactile learner, a sequential or global learner, or a deductive or inductive learner, this chapter offers you suggestions for enhancing your learning.

From here, the path leads directly into Chapter 8, where we examine useful math study skills and winning strategies for learning math and reading math textbooks.

Chapter 9, one of the most important in this road map, teaches you how to deal with test anxiety and how to perform your best on math exams. It reviews test preparation and test-taking strategies as well as good nutritional guidelines to follow before exams. This chapter contains a powerful visualization exercise to help you feel alert, clear, calm, confident, and competent on exams. You will also find this exercise recorded on the accompanying CD.

Chapter 10 explores problem-solving approaches and ways to "think like a mathematician." We end the journey with Chapter 11, a look at your exciting math future.

A special music and dialogue CD accompanies this workbook. It was produced to help you experience some of the wonderful calming and visualization techniques described in this workbook. The Calming Breath and the Deep Abdominal Breath both focus on calming your mind and body. The "Success in Math Visualization" will help you reprogram yourself to succeed in math, and the "Math Test Anxiety Reduction Visualization" will take you through a powerful technique to help you achieve success on math exams.

Some Basic Axioms

In this workbook, I present a comprehensive, well-tested plan for tackling your math fears and anxieties. My approach is based on the following premises:

- As a student, you need not be forever burdened by the negative experiences or unproductive messages from your past.
- You can learn to manage stress and anxiety physiologically so they can be productive rather than destructive aspects of your academic performance.
- By changing negative math self-talk to positive self-talk, you can greatly improve your ability to deal with math.
- Learning how to maintain a positive attitude toward math and toward your ability to do math has a tenfold beneficial effect on your math performance.
- By using visualization techniques, you can reprogram yourself to succeed in math.
- You can increase your ability to learn math by using more appropriate, effective, and efficient study skills and learning-style strategies.
- Conquering math test anxiety is like winning the battle.
- Math can actually begin to be fun and exciting for you!

Guidelines for Using This Workbook

Here are a few suggestions to help you maximize the benefits of the exercises in this workbook:

1. *Give yourself some uninterrupted time.* Set aside a special time and place where you will not be distracted or interrupted. Even if you have only 5 or 10 minutes, make sure it is a private time just for you. Unplug the phone if necessary. Put a note on your bedroom door saying you can't be disturbed for a few minutes.
2. *Work alone and in silence.* You will be able to gain greater self-understanding and insights if you work alone and don't prematurely share your new perceptions with others. There will

be plenty of time later, after completing your work, for you to discuss what you've learned. I encourage you to share your insights with supportive people, such as friends, teachers, or a counselor. They may make your journey easier.

3. *Keep a math journal.* You will probably find it helpful to keep a journal in which you write about the new awareness, sensitivities, and insights you gain. This may be similar to Exercise 2-3 (Chapter 2), or it may be in the form of a traditional diary.

4. *Assume a comfortable position.* Sit quietly, with your eyes closed, in a slightly darkened room when listening to the CD or taking yourself through the anxiety management exercises in Chapter 3, the Wall Fantasy in Chapter 4, the "Success in Math Visualization" in Chapter 6, and the "Math Test Anxiety Reduction Visualization" exercise in Chapter 9. Be sure to arrange to be alone and undisturbed so you attain the full benefit of these experiences.

5. *Be persistent.* The road to success can be achieved only through staying power, resolve, and determination. "Stick-to-it-tiveness" will help you reach your math goal. You must master many steppingstones along the way: from learning how to reduce anxiety to overcoming psychological stumbling blocks; from rewriting disempowering math beliefs to reprogramming yourself to succeed in math; from learning to use effective math study skills to conquering math test anxiety. It will take time and patience. So persevere, stay with it, and you'll conquer your math fears and anxieties.

6. *Be positive.* Know in your heart and mind that **you can and will succeed.** Have faith that you are able to achieve in life what you realistically desire and work toward. Your persistent, positive efforts can and will be rewarded. I've seen it happen for others who have conquered math anxiety. Why not you?

7. *Jump in and start anywhere.* Although I have written this workbook as a detailed road map to help you reach your math success goal, you need not progress through the workbook in sequence. As much as possible, each chapter stands independently of the others. Thus, you may wish to work mainly on those areas that concern you the most.

Summary

Conquering Math Anxiety: A Self-Help Workbook offers you a detailed road map laying out all the steps along the route to overcome math anxiety and ultimately to achieve success in math. Many students have taken this route before you and have succeeded. So can you!

Do You Really Have Math Anxiety?

Do not worry too much about your difficulties in mathematics. I can assure you mine are still greater. —Albert Einstein

Labeling yourself "math anxious" is almost like having "medical student's disease," a common phenomenon among medical students. They diagnose themselves with a devastating disease based on having even one manifestation of the disorder.

Not unlike a disease, math anxiety has clear-cut symptoms. These symptoms include negative emotional, mental, and/or physical reactions to mathematical thought processes and problem solving caused by discomforting or unrewarding life experiences with math.

The anxiety, fear, and blocking behavior you have with math may not be primarily due to math. Although these reactions appear to arise in your math class or on a math test, they may not be caused by math itself. Math acts like a fine magnifying lens, bringing into sharp focus a host of other academic deficiencies, like poor study skills, knowledge gaps, or inadequate test preparations or test-taking skills. The anxiety you express about math thus becomes a symptom and not the disease itself. So you may not actually have "the math anxiety disease," although you are experiencing some of its symptoms.

Often, students who believe they are math anxious in reality are merely victims of test anxiety. Jonathan, an 18-year-old, is a perfect example of this. He sought counseling from me, distraught because he was failing math but felt he really knew the material. He was sure his poor performance was due to a math block or perhaps a

deep-seated fear of math. Upon questioning Jonathan, it was obvious he did know his math. But when he went into an exam, he would totally "blank out." Sometimes, it was as if he couldn't even add two numbers together without making a mistake. I learned he occasionally blanked out or panicked on exams in other subjects as well. Halfway through a chemistry exam, he panicked and began making all sorts of errors on problems he really knew how to answer. He also found himself anxious and perspiring a lot during his psychology final and ended up ruining his chances for an A in the course.

If you are like Jonathan, panicking on math tests as well as on exams in other subjects, the help you need is primarily in the area of overcoming test anxiety. A reduction in the high levels of generalized anxiety you experience when taking tests will increase your ability to perform, not only in math but in all your courses. Chapter 9 in this workbook, entitled "Conquering Test Anxiety," is an excellent chapter to study. It will give you tried-and-true strategies to prepare for tests, take tests, deal with the anxiety, and use visualization techniques to ensure successful results.

I have also worked with other students who were not truly math anxious but whose academic difficulties severely affected their ability to succeed in math. Perhaps you are one of these students. For example, you could easily find yourself feeling overwhelmed, anxious, and "over your head" in a math course if, unknowingly, you missed important preparatory coursework along the way. You may be surprised to learn your anxiety in math might abruptly end if only you were properly placed in an appropriate math class or were to receive remedial help in your deficient areas. I have worked with many students who experienced severe anxieties in college algebra because of sizable knowledge gaps in their math background, despite the fact they never skipped a semester of beginning or intermediate algebra. Upon further examination, I learned, although these students had never actually missed any classes, they struggled with basic algebraic concepts like graphing a line or solving fractional and quadratic equations. For these students, it was like taking German 4 when they had missed learning how to conjugate verbs in German 2. They were simply ill prepared for their current course. They had good reason to be anxious.

If you have missed information along the way, it's almost impossible to expect to clearly understand any subject. Learning math is very similar to learning a foreign language, with new terms and meanings for words and letters, but it is also similar in that the early courses are designed to provide the grammar and syntax for later courses. Each course builds upon the previous ones. Early math courses also provide the "grammar and syntax" of what is to come later—that is, it's the "language" of math. In math, in particular, each step is an essential steppingstone for the next. When constructing a large, impressive skyscraper, we must be sure the foundation is strong and sturdy. And as we go higher and higher in math, it is like building a tall, elegant structure. If you have missed important knowledge along the way, you may have to go back, find those missing blocks, and reconstruct your building. Once you do this and your building stands on a solid foundation, you'll find math really can become both rewarding and fun.

Fine, you say, but how do you accomplish this? There are a number of options open to you. If the gaps in your knowledge are substantial, you should consider actually repeating or, if possible, auditing some of the math courses you have already taken in the past. Many community colleges offer developmental math courses, such as the fundamentals of math and beginning algebra. Also, often available are courses that are modularized or separated into smaller units so that you can elect to take only the necessary unit(s) to fill in your background. I've met students who were deficient in some specific information taught in the latter part of Algebra 1 and who were delighted to learn they were able to register for just the third module of an Algebra 1 course at our college.

Ellen is another student who wasn't truly math anxious but whose academic difficulties affected her ability to do math. Ellen understood her math when it was presented in class, but when she sat down to do her homework 2 days later, it reminded her of hieroglyphics, mysterious and undecipherable. She fell behind in her class work, and soon she became distraught with math. At this point, Ellen came to me, convinced she was math anxious. When she described how she studied math, that she failed to review her math immediately after class and didn't make up her own practice exams (among other things), I was sure Ellen's problems were related to poor study and test-preparation skills and not to math anxiety. I spent several sessions with Ellen and taught her good study strategies. Soon Ellen's grades—and her comfort level—began to improve. If your discomfort in math is related to poor study skills or inadequate test preparation, Chapters 8 and 9 in this book can help you.

Goals for Success

Before we continue, let's be very clear about what you want to achieve. What are your goals for success in math? If you're taking a math class now, what do you want to get out of it? What does success in math mean to you? Goals are like anchors to the future we desire. We must toss them out in front of us and then use them to pull ourselves along. In this way, you take control of your math future. Your math success will happen by your design and not by chance. You *can* have the math success you want!

In Exercise 1-1, I'd like you to identify what you wish to achieve in math. You may have a short- or long-term goal. Write down your goal and be specific. For example, you might have a short-term goal such as: "I want to solve quadratic equations," or "I want to learn how to work out the word problems in my math book," or "I want to successfully complete my Calculus 1 course this semester." Long-term goals might be: "I want to progress through my algebra courses and through calculus at a reasonable pace over the next 2 years" or "I want to enroll in a math course next fall and complete it with at least a C grade." It's important for you to be certain, without a doubt, this is the goal you wish to achieve and to present it without alternatives. The less conflicted you are about achieving your goal, the greater the probability you'll accomplish it.

Next, put down a target date for when you plan to reach your math success goal. This is the date you think you realistically can achieve your goal. Some students find it's best to set three dates, one being the most optimistic date, one being a more moderate and realistic date, and the third being the latest acceptable date.

The third part of the exercise asks you to evaluate the strength of your goal to help you to determine how strongly you are motivated to achieve it and what its value is to you.

In the fourth part of this plan, establish the benefits you will get when you reach your goal. List both the tangible as well as intangible benefits; sometimes, the intangible ones are more important than the tangible ones. For example, Remy wrote that success in reaching her math goal would make her feel better about herself and help her to know, one day, she could be successful as an architect. For Carl, success in math would prove to himself his seventh-grade teacher was totally wrong when he said Carl would never be able to do well in math.

In the fifth part of the exercise, identify some of the obstacles you may face along your path, as well as steps you may be able to take to overcome these obstacles. Miguel, a college freshman, identified one of the major barriers to his success in math: poor study skills. He decided to research the best study strategies for improving his math and to take a special math study-skills workshop offered at his college. Emily's major obstacle was her own negative and embarrassing past history with math, which loomed over her like a large, dark rain cloud. She knew this stopped her progress and she had to take some drastic steps to deal with it. Emily decided she needed to seek counseling to work on her poor self-esteem.

In part six of the exercise, look at the barriers and the possible steps required to overcome these barriers; then discern the positive forces and abilities you can use or strengthen to meet your goal. Jasmine noted in her plan that she has lots of determination, she is motivated, and she is willing to work hard. Chan felt his good study habits and ability to overcome obstacles were his biggest assets.

Next, specify the supportive people in your life who can help you on the road to success in math. Who are these people, and how will they be able to help? Perhaps you have a teacher who encourages you, a friend who is a great math tutor, or a companion who is particularly reassuring.

In the eighth section of this plan, I ask you to select the significant action steps you are ready to take to meet your math success goal. Include here the measures you think will work best. For example, you may explore the availability of math tutorial programs or decide to work on a specific number of math problems at each study session. Many of these steps you will be aware of only as you continue to read further in this book. As you read, be sure to occasionally return to this list and add relevant steps to it to further personalize this plan for you. For each action step, list a specific target date by which you will accomplish it.

The last section of this plan asks you to determine how you will reward yourself for meeting your goal. The reward is a very important part of this plan. It will help keep your motivation high and sustain your effort along the path to success. I've seen all types of rewards described on finished plans, from a night out on the town to a trip to a favorite vacation spot to buying a special gift for yourself. Remember to give yourself a reward at the successful accomplishment of your goal. You deserve it!

EXERCISE 1-1 **My Plan for Math Success**

1. I list one realistic math success goal I wish to achieve. I state it in specific, positive, measurable terms. I write out, vividly and in detail, exactly what I want. My math goal is:

_____

_____

2. The realistic target date for achieving this goal is:

3. My math goal should meet the following goal-setting criteria (check all that apply):

_____ I have clearly stated it.

_____ I value it.

_____ I believe I can do it.

_____ I want to do it, and I am motivated.

_____ I find it rewarding and personally fulfilling.

_____ I am clear that this is what I want, as opposed to other choices.

_____ It is a realistic possibility for me in terms of my time and ability.

_____ I envision a plan of action for achieving it.

4. I want to achieve my math goal because of the following benefits and potential satisfactions (list as many as possible; include both tangible and intangible benefits):

5. These are some barriers or obstacles I may face and steps I will take to overcome them:

Barriers   **Steps to Overcome Barriers**

6. These are the positive forces and abilities I can use or strengthen to meet my math goal:

7. These are the people who can help me in achieving my goal:

Name   **Type of Help They Can Give**

8. The significant *action steps* I need to take to meet my math success goals are:

Action Steps   **Target Dates**

a. _____ _____

b. _____ _____

c. _____ _____

d. _____ _____

e. _____ _____

f. _____ _____

g. _____ _____

9. Here's how I will reward myself for meeting my math goal:

One of my students, Mary Jo, developed a wonderful math success plan. Mary Jo wants to become a neurosurgeon someday, so completing her math goal was very important to her. Here is her plan:

1. I list one realistic math success goal I wish to achieve. I state it in specific, positive, measurable terms. I write out, vividly and in detail, exactly what I want. My math goal is:

I want to pass intermediate algebra this semester with at least a C grade, with the help of my teacher, my counselors, the tutors, my support system, and my willingness to do it!

2. The realistic target date for achieving this goal is:

the end of this semester

3. My math goal should meet the following goal-setting criteria (check all that apply):

___X___ I have clearly stated it.

___X___ I value it.

___X___ I believe I can do it.

___X___ I want to do it, and I am motivated.

___X___ I find it rewarding and personally fulfilling.

___X___ I am clear that this is what I want, as opposed to other choices.

___X___ It is a realistic possibility for me in terms of my time and ability.

___X___ I envision a plan of action for achieving it.

4. I want to achieve my math goal because of the following benefits and potential satisfactions (list as many as possible; include both tangible and intangible benefits):

My tangible benefits are:

1. *Learning math would enable me to meet my career goal of being a doctor.*
2. *I would be able to get higher-paying jobs.*
3. *I would be able to help my friends and kid brother with math home-work.*
4. *It would help me to be successful in my science courses.*

My intangible benefits are:

1. *I would have more confidence to take college algebra and calculus.*
2. *I would have improved self-esteem.*
3. *I would feel good about my accomplishment.*
4. *I would be more relaxed dealing with math in my daily life.*
5. *I would prove to myself that I can be successful as a premed major.*

5. These are some barriers or obstacles I may face and steps I will take to overcome them:

Barriers	Steps to Overcome Barriers
a. fear or panic	a. breathing exercises or self-talk to calm my nerves
b. old tapes of my parents telling me I don't have to do well in math since my mom didn't	b. telling myself I need to know and do well because I am not my mom
c. being a perfectionist	c. emotionally accepting what I already know intellectually: Nothing in life is perfect
d. perceived lack of time	d. I must set aside specific blocks of time to study math.

6. These are the positive forces and abilities I can use or strengthen to meet my math goal:

   My willingness and desire to learn math.

   I'm not afraid to raise my hand and ask questions in class.

   I have perseverance and patience.

   Like the mountain climber—once I set out, I am determined and I don't give up until I get where I want to go.

7. These are the people who can help me in achieving my goal:

Name	Type of Help They Can Give
Ms. Collins, my math teacher	She can clarify topics I'm confused about.
Mr. Soto, my counselor	He offers me emotional support.
Iliana, Hiro, and Sasha, the math lab tutors	They can review my math step by step.

8. The significant *action steps* I need to take to meet my math success goals are:

Action Steps	Target Dates
a. Do all my homework as soon as I leave class	*every day*
b. Ask my teacher for extra help	*whenever needed*
c. Go to the tutors regularly	*twice weekly*
d. Arrange a quiet time to study math	*every day*
e. Ask my older sister for emotional support	*by Sept. 20*
f. Make up practice tests for myself	*once a week*
g. Do relaxation and visualization before my exam	*Sept. 21, 22, 23*

9. Here's how I will reward myself for meeting my math goal:

I will take a trip and visit my best friend in Los Angeles and we will go to Disneyland.

EXERCISE 1-2 **Collaborative Learning: Share Your Math Goals and Action Steps**

With a small group of classmates, share your plan for math success. Make sure to share your action steps and how you'll reward yourself for meeting your goal. In the space provided, record any helpful ideas you learn.

_____

_____

_____

_____

_____

Summary

In this section of our road map, I have asked: Do you really have math anxiety? We've seen that much of the fear and discomfort associated with math is often the result of a host of other academic difficulties. You were asked: What does successful accomplishment in math mean to you? I've encouraged you to explore your goals for math success and to develop a plan for achieving them.

The Math Anxiety Process

Juanita has been math anxious as far back as she can remember. She hates math. She dreads it. She avoids any contact she could possibly have with numbers. Where did it all begin, I wondered. I asked Juanita to write her math autobiography so we could get some insight into the roots of her problem. And there it was. The mystery of her great fears lay in front of me in black and white. Juanita wrote:

> I remember I was in grade school, and I loved it. But the nuns were very strict. And one day I had to go to the bathroom real bad. I raised my hand to ask, but the teacher didn't wait for me to ask for permission to leave; instead I was called to the chalkboard to complete a math problem. And there at the board I lost control of my bladder. It was awful. Right in front of the whole class. I've hated math ever since.

Ousmane's math autobiography revealed his math anxiety didn't begin until middle school. He was taking a prealgebra class and having difficulty understanding several of the concepts. He wrote:

> My teacher became so angry with me, he yelled at me in class, saying I'd never be able to do math. And he was right. Ever since that time math has been my worst subject. I'd avoid it now if I could, but I need it for my major.

John remembered needing help with his math homework in third grade. His older brother began to help him and tried to explain some of the concepts. When John still had difficulty understanding these concepts, his older brother beat him up. John made up his mind at the time never to learn math again. He was 45 years old when he came to me and said, "I'm ready to begin learning math."

Rachel had a lot of difficulty in her fifth-grade math class. She wrote in her math autobiography:

> My teacher was so frustrated with my asking so many
> questions in class that every time she taught math she
> had me sit in the hallway outside of class. She said I
> was a hopeless case and I couldn't learn math anyway.

As unbelievable as these cases sound, they are all true examples of students who have come to me for math anxiety counseling. The negative situations they encountered were very painful and discouraging. These students, and perhaps many of you, have had the misfortune to encounter unkind people who do not help foster positive math feelings. These people did not act in your best interest. They were unkind to blossoming math skills. They are not the best teachers and guides. They often perpetuate a negative and untrue belief about your ability to learn math. But despite it all, you persist. You continue on your journey in the noble pursuit of learning math. You, and only you, have the power to change your life.

The techniques I describe in this workbook and on the CD have helped these students overcome their fears and successfully accomplish their math goals. My strong message to them—and to you—is that *your past negative math experiences needn't continue to burden you*. I encourage you to throw off the shackles of the past. Starting today, tell yourself, "I can and will succeed in math."

Reexamine Your Past Math History

I've been able to identify several different reasons for the onset of math anxiety in the students with whom I've worked. Let's explore some of these reasons and see how they might have affected you.

Embarrassments

Have you ever been embarrassed about math? Did you feel shy or mortified when being asked to do math in front of others in class? Many, many students report embarrassing moments related to math, dating as far back as first grade all the way through high school. They quivered with self-consciousness when called upon to answer a math problem or do their times-table in front of the class. One wrong answer and they felt humiliated. Many feared being called to the chalkboard to work out a math problem, standing there all alone, with everyone watching. Public speaking fears and stage fright became associated with embarrassment over math. Other students reported being drilled over and over with flashcards, and how flustered they became when they forgot even the easiest of facts or when they were caught counting on their fingers. Still other students were mercilessly teased by classmates if the answers came too easily and they seemed to enjoy math. They were called names like "nerd," "show-off," or "geek."

Negative Life Experiences Associated with Learning Math

Many students have reported traumatic childhood experiences quite unrelated to math, but somehow the emotional upset of these experiences became associated with math. Phyllis's parents got divorced

"NO, NO, NO WRONG AGAIN!"

ANOTHER CAUSE OF MATH ANXIETY

when she was learning long division, and to this day, Phyllis is distraught whenever she is required to do math, especially division. Francisco's family suffered a lot of emotional turmoil throughout his middle school and high school years. His father was an alcoholic, and family life was punctuated by many arguments. His home situation severely affected his ability to concentrate on math during those years, and math still remains an area of extreme discomfort to him. Norm's grandmother died when he was in sixth grade. She had been living with his family, and her absence was taken very badly by everyone, especially Norm. He began to fall farther and farther behind in math and did poorly on his math tests. Soon Norm began dreading the thought of ever having to take a math class. He would do anything to avoid math.

When you look back to the days when you first began to have math discomfort or fears, can you identify events in your life that were so emotionally disturbing they could have become associated with learning math? Or perhaps you experienced an illness or other interruption in your education that caused a critical gap in your math background. Many students also report a lack of family support when faced with math difficulties (Trujillo & Hadfield, 1999).

A. AREM

Social Pressures and Expectations

Did family members or friends ever try to tutor you? Did they ever reprimand or show disgust when you didn't seem to comprehend math quickly enough? Did your parents push you to succeed or compare you to a sibling who was a "math whiz"? Or perhaps you had relatives or classmates who discouraged your math achievement, telling you they were never good in math either. I've worked with students who had nightmares as a result of being tutored by demanding or overzealous relatives, and I've met others who vowed never to take math because they hated being compared to a sister or brother who excelled at math.

Margo's father was an engineer who very much wanted Margo to be as competent in math as he was. Every night, he tutored her, went over her homework, and tested her on her progress. Margo wanted so much to please her father, but when she didn't understand the math fast enough, he often got so frustrated he slammed the book down and walked away. By the time she got to high school, she began to avoid math. Unwilling to repeatedly face her father's disappointment in her, Margo decided to take as little math as possible.

It has been well established that parents' and teachers' expectations influence children's math achievement and their beliefs about their ability to learn math (Brannon, 2008; Eccles & Jacobs, 1986; Yee & Eccles, 1988). In addition, it's been found social support is related to both the choice to take advanced math and the grades achieved in math. One study showed a strong relationship between the support students received from their parents, teachers, and peers and the choice to take advanced math courses in high school (Sells, 1980). For example, girls supported in taking advanced math courses were more likely to earn As or Bs, but those without such support were likely to get Cs or Ds.

Desires to Be Perfect

Have you feared math because it seemed so exact to you that there always appeared to be a right and a wrong answer? And if you got a wrong answer, did you feel it was a poor reflection on you and your academic abilities? Many students feel like this. They often have a fragile, negative self-image. They don't feel good about themselves,

and they try to avoid situations that point out their weaknesses. Math puts them in such a situation. Each time in the past when they were called on in class and gave a wrong answer, it made them feel less worthy, dumb, or unintelligent. They would try and try to work math problems, but when their answers came out wrong, they were left with feelings of incompetence and poor self-esteem. Is it any wonder they began to avoid math? Has math affected you this way?

Poor Teaching Methods

In elementary or middle school, did you have teachers who really didn't like math or want to teach it or who weren't well trained to teach math? Did you have teachers who couldn't answer your questions or perhaps would put you down when you didn't know the answer to a question? Did you find your teacher couldn't quite explain the material in a way you could understand?

One student reported his fifth-grade teacher said if the class was good all day, they wouldn't have to do their math that day. Math became a punishment for the class if they misbehaved. It turned out this teacher was trained as a drama teacher and was quite frightened by math. In the past, many college students who were math anxious chose elementary school teaching to avoid most college math requirements. In doing so, they successfully sidestepped any confrontation with their anxiety, but they still harbored it. Since math is taught in all the elementary grades, these teachers had ample opportunity to pass their math fears and uncertainties onto their impressionable students.

In other situations, teachers have ridiculed students or told them they would never be able to learn math. Teachers rushed through the material or gave few or no problems for homework to reinforce the concepts presented. In some cases, students needed to have hands-on math experiences, and the teacher only lectured or wrote on the board but couldn't provide adequate alternative teaching methods to accommodate kinesthetic/tactile learning styles. Other teachers weren't trained in or did not implement effective teaching methods such as collaborative and cooperative learning groups to help students learn math.

In one study, researchers set out to trace the roots of math anxiety by conducting in-depth interviews with 50 elementary education majors in elementary-mathematics-methods classes (Trujillo & Hadfield, 1999). The results of this study were astonishing. All the students in the study had had several negative classroom math experiences as well as a lack of parental support at home. In addition, all students suffered from severe math anxiety and math test anxiety, and they were fearful about teaching mathematics themselves. Yet, despite their own negative experiences with math, these students planned "to employ constructivist and developmental methods they learned in their college mathematics methods classes in order to make mathematics more meaningful to their own students" (p. 9). These future elementary school teachers were determined to provide a more positive math learning environment for their students than they had experienced. The authors concluded that negative classroom math experiences, combined with lack of parent support and fear of math tests, were major contributors to math anxiety.

In another study, researchers examined the role of instructors' behaviors in creating or exacerbating math anxiety in students from kindergarten through college. A total of 157 education majors in a senior-level elementary-mathematics class were asked to respond to the following statement: "Describe your worst or most challenging mathematics classroom experience from kindergarten through college" (Jackson & Leffingwell, 1999).

Surprisingly, only 11 of the 157 students reported having had only positive math experiences from kindergarten through college. Of the remaining 146 students, all underwent some traumatic math experience. Although some students were traumatized as early as kindergarten or first grade, the researchers found three main grade level clusters in which there were initial traumatic math encounters: (a) in the third or fourth grades, (b) ninth through eleventh grades, or (c) during their college freshman year.

The instructor behaviors most cited as disturbing to the students were scowling or having a demeaning manner; seeming angry, hostile, uncaring, or insensitive; making derogatory comments; and having a gender bias. Also cited were teachers giving poor quality of instruction, poor explanations, or rushing through materials.

In addition to poor teaching methods, sometimes curriculum choices have adversely affected students. These include unsatisfactory textbook selection, inadequate prealgebra preparation courses, gaps in course or unit sequencing, too fast a pace, limited review or practice sessions, and materials or approaches incompatible with students' learning styles.

Negative Math Games People Play

Do you ever say things to yourself that block your ability to do math? Your internal mind talk can play an important role in your math performance and could easily be a negative or destructive influence on you. When this talk is detrimental, we refer to it as *negative math games*. These games, if not reversed, can result in a complete loss of self-confidence in math, and self-confidence is one of the most important aspects of achieving math success. The following statements are examples of self-defeatist games students play on themselves. You may say things like: "I was never good in math, so I can't be good now." "Why do I need math anyway?" "Math isn't useful in my life." "I don't work math quickly enough." "Everyone knows how to do it but me." "This is a stupid question, but . . ." "It's too easy; I must have done it wrong." "I got the right answer, but I don't know what I'm doing."

Aside from the negative math games you play on yourself, there are games others play on you. These are statements others say to you, deliberately or perhaps unintentionally, that negatively affect your ability to do math. Perhaps you sought help on a math problem and the other person responded: "Oh, that's easy" or "You should know that by now." "You'll never be able to do math." "You'll just have to work harder in math, and you'll get it." "The answer's right in front of you; don't you see it [you dummy]?" Statements like these just serve to make you feel bad and to increase your fears and uncertainties about math. Other games people use include, "You got the right answer, but you did it the wrong way." Since there are many ways to solve a math problem, this game serves to increase both your self-doubt and your anxiety level. Still others might try to discourage you from learning math by saying, "Why learn math anyway? You'll never need it."

Cultural Myths

Have you accepted the powerful myths surrounding math? Our society too often accepts mathematics illiteracy. Few people would admit they can't read. However, those who cannot do math will find lots of company, and acknowledging their deficiency not only produces no social stigma but generates empathy and commiseration. Furthermore, the popular media tend to portray those skilled in math as intellectually superior and, therefore, strange or different.

Throughout our society, parents, teachers, friends, relatives, books, magazines, and television have often perpetuated a system of false beliefs about math. It is a belief in these falsehoods, which have no basis in reality, that can stop you from making progress in math. Let's look at some of these beliefs now.

Do you believe that you must have a "math mind" or be a mental giant to succeed in math? Whereas there are people who have extraordinary aptitude and ability in math, as in any subject, you needn't be a genius to become competent. You *are* able to do math. There is no special innate ability we inherit that relates only to this subject. We are all capable of learning math. It takes time, patience, determination, and a great deal of practice.

Are you convinced there is a magic key to doing math, or math problems must be worked on intensely until they are solved? Do you think there is a best way to solve each math problem? Once again, all these are falsehoods. There is no magic key or formula for doing math. There is no one best way to solve any problem. Usually, there are a variety of different ways to find the correct answer. And math problems often are not solved in one sitting. It is important to take breaks and rest during problem solving. Mathematicians may take several days, and sometimes months, to solve a difficult problem.

Have you often thought you should be able to do math quickly in your head and only dunces count on their fingers? Any physical model that helps you solve math problems is okay. For example, finger counting may actually show you understand what you're doing and you aren't merely doing math through memorization. As for doing a math problem in your head, this is too much to expect from anyone who hasn't done this sort of thing many times before. Even experienced mathematicians can't necessarily work out new math problems in their heads.

Have you thought math is not creative or math requires only logic and rational thinking but not intuition? Perhaps you've believed you must always know how you got your answer or that it was always important to get the right answer in math. None of these statements represents reality; all are math myths. Math can be very creative, imaginative, and intuitive. Mathematicians often use their intuition to figure out solutions, and they can't always explain how they arrived at their answers. Problem solving can be a very creative, innovative process. To play with various methods, test out what feels right, sleep on the problem, or brainstorm different solutions can all be fun and inventive. Knowing the precise answers often doesn't matter because many times answers to difficult problems may only be approximations or "guesstimates."

Here, where we reach the sphere of mathematics, we are among processes which seem to some the most inhuman of all human activities and the most remote from poetry. Yet it is here that the artist has the fullest scope of his imagination. —Havelock Ellis, *The Dance of Life* (1923)

Gender Stereotyping and Socialization

Do you believe that only males are good in math or that careers requiring math, such as engineering, technology, or science, are mainly for men? The math-anxious males with whom I've worked would certainly disagree. They feel men aren't any better in math than women.

Studies show gender-related differences in math performance are not present in the general population despite the stereotype that females dislike math and do poorly in it. This stereotype is quite destructive and often leads to biased treatment toward females from teachers, parents, and peers (Brannon, 2008).

Throughout elementary school, girls perform similarly to boys or demonstrate higher proficiency in arithmetic computational skills (Fennema, 1980; Hyde, Fennema, & Lamon, 1990). Things start to change for girls around age 12. Studies show they begin to feel less confident than boys about their math ability, and this trend, along with believing math isn't important to them, increases into adulthood (Eccles, 1989; Kimball, 1995). Math becomes seen as a male domain (Kimball, 1995). It isn't surprising, then, that we find high school girls often choose to take fewer math electives than boys, and their performance on standardized math tests is lower than boys'. Taking fewer high school math courses has long-lasting negative consequences for females. Math acts as a "critical filter," barring those who are deficient from science, engineering, computer, and technology careers (Hyde, Fennema, & Lamon, 1990; Sells, 1980).

Cross-cultural research hasn't substantiated a universal pattern of males outperforming females on standardized math tests. In some countries, female students' achievement on math tests is equal to or superior to males'. This highlights the fact that American females' poor representation in fields requiring math is most likely a cultural phenomenon (Feingold, 1988, 1994; Hanna, 1989; Huang, 1993; Lummis & Stevenson, 1990; Skaalvik & Rankin, 1994).

In another interesting study, researchers evaluated 220 female students' math performance after they were given a bogus scientific explanation for alleged gender differences in math performance. The students were told one of four possible explanations: (a) There is a genetic (or inborn) reason for women underperforming in math. (b) There's an experiential reason for women underperforming, such as boys receiving preferential treatment. (c) There's a common cultural stereotype that females underachieve in math. (d) There are no sex differences in math. The researchers found that females given the genetic explanation for women's underachievement had the worst math performance. Comparable low performance was also found in those reminded of the common cultural stereotype. The good news is that those students who received the experiential explanation or were told there was no difference both performed much better (Dar-Nimrod & Heine, 2006).

In one study of female college students, researchers found that although women may explicitly disavow stereotypes and report egalitarian values, they could still be influenced by negative stereotypes on the preconscious or "implicit" level. In other words, even if they consciously think that women can perform just as well as men in math, on a preconscious level they may still associate math achievement with males and not with females. The researchers noted, "The less women were gender identified and the less they possessed implicit gender stereotypes about math ability, the better they performed on the final exam and the more likely they were to express interest in a math-related career" (Kiefer & Sekaquaptewa, 2007, p. 16).

In a monumental study, psychology Professor Janet Hyde and her colleagues looked at 7 million math scores from ten state standardized exams to determine whether gender differences exist (Hyde et al., 2008). They found that average math scores were the same for both genders. Even at the top scoring percentiles, there was little difference.

Science Daily, an online research news journal, notes that "contributing to the public's notion that boys truly are better at math" is the fact that males often score better on the SATs than females do (2008, p. 2). With regard to the SATs, Hyde and her colleagues explain that many more females than males now take the SATs because a greater proportion of females than males attend college. As a result, the male scores represent more of the top talent among the male student population, while the female SAT scores represent both the top and lower tiers of female talent.

So it is sad but true that with all these studies indicating that females are capable of performing at equal levels to males in math, negative stereotypes about female's math performance continue to exist.

Dyscalculia: A Cause or a Symptom?

For a small percentage of students, dyscalculia may be the cause of math anxiety. Yet, for others, math anxiety may manifest itself as dyscalculia or symptoms of innumeracy (Ashcraft, 1995).

According to the National Center for Learning Disabilities, "dyscalculia is a term referring to a wide range of life-long learning disabilities involving math" (NCLD, 2006). Studies have indicated that approximately 4% of students experience dyscalculia (Sharma, 2003).

Dyscalculia usually first shows up in elementary school, when a child has difficulty solving basic arithmetic problems and shows a poor understanding of number concepts and the number system. The student often is unable to recognize patterns when adding, subtracting, multiplying, or dividing and does not understand what arithmetical operation is needed for a problem. Other difficulties have also been noted, such as when the individual is "counting, giving and receiving change, tipping, learning abstract concepts of time and direction, telling and keeping track of time, and the sequence of past and future events" (Vaidya, 2004).

The causes of dyscalculia are not clear, but research points to a myriad of possible causes. These include heredity; a biological malady within the brain; emotional problems due to environmental deprivation; low intelligence; math anxiety; ineffective teaching strategies used during early childhood; or poor social skills (Michaelson, 2007).

If you have been diagnosed with dyscalculia, check out the following Web sites for information: www.dyscalculia.org, www.ldanatl.org, and www.ldonline.org.

The anxiety reduction exercises described in Chapter 3 and the exercises on building positive self-confidence in Chapter 5 will help. In addition, see the section entitled, "Helpful Strategies for Dealing with Dyscalculia," in Chapter 8.

What Is Your Math History?

So, as you can see, the origins of math anxiety are varied. The following exercise will help you focus on your math history and will help you identify the early roots of your problem.

EXERCISE 2-1 **Your Math Autobiography**

What is your math history? In the spaces that follow, briefly describe your chronological history in terms of the negative and positive experiences you've had with math. Include your earliest memories, as well as memories of how your teachers and your family influenced you in math. Describe how your family members approached math and describe their attitude toward your math ability. Include a description of how you've dealt with recent situations involving math in other classes, on the job, or in daily life situations. End with a discussion of how math could help you to accomplish your educational objectives, earn more money, choose a career, or in any other aspect of your life. (Use additional paper.)

_____

_____

_____

_____

_____

_____

_____

_____

_____

_____

Here are excerpts of math autobiographies written by students who are working on overcoming their math anxiety. Perhaps you can identify with one of them.

Theria, a junior, who hopes to become a Web page designer, wrote:

I remember being very good in second-grade math, but I was never praised by my teacher or my parents. In third grade I noticed the boys in my class were always selected for math contests and games. I thought they must be really good at math and it wasn't important for me to excel. By sixth grade, my grades in math declined. My father tried to tutor me, but he yelled and didn't have the patience or the knowledge to help me through my math deficiencies. My mother justified my failures by telling me she was never good in math either. "It's no big deal; I turned out fine, didn't I?" So now I had a justification not to try. My father frightened me and my mom let me know that failing math was okay.

Ricardo, a tall, lanky boy, who hopes to become an astronomer one day, wrote:

At a young age I found math fun and challenging. It came very easily to me (aside from the fact that I didn't know my times-tables and to this day I still count on my fingers). Throughout the early years in school, I received As in math. I even did all right in prealgebra, ending the year with a B. But when it came to algebra, things started to get more difficult. In my other high school classes, I was able to "float" through the homework, many times doing the assignments right in class. But math was different. It took discipline that I did not have. I didn't do my algebra homework and my grade suffered as the result of it. I started getting Ds.

Jefferson, a member of the school basketball team, wrote:

From my earliest memories in elementary school, math was something to fear and despise. I was always the last to finish and the most likely to fail an exam. As a result of my poor performance and not wanting to be left behind, I became a professional cheater. This habit of mine continued until I reached high school, where I was put into prealgebra. I found myself way over my head in a class with students far more advanced than I. Every homework problem I did was wrong and every exam was a failure. Cheating didn't help. My instructors in high school tried to help, but due to the large class sizes, they couldn't tutor me. Soon after this, I began to skip class and became very indifferent toward school.

Reassess Your Past

I want to reiterate here what I stated earlier: ***Your past negative math experiences need not continue to burden you.*** It is both your responsibility and within your power to take control of your math destiny. In the next exercise, I encourage you to review your math history so you can understand the roots of your fears. Then I will ask you some important, thought-provoking questions to help you reevaluate your previous experiences. You can throw off the shackles of the past that have weighed you down so you can make progress.

EXERCISE 2-2   Collaborative Learning: New Insights and Revelations

When you finish your math autobiography in Exercise 2-1, look it over and see if you can find the roots of your math anxiety. Is it clear where it began? What factors most influenced you? What past attitudes and actions on your part contributed to your current difficulties? Then ask yourself: "Are these conditions still relevant in my life today? Are the messages I carry around about my abilities to do math true today? Can I move further and not let the problems of the past be my perpetual stumbling blocks to learning math? What positive experiences can I build on?" In the space provided, describe the insights you have gained from reading over and analyzing your math history. Share your insights with a small group of classmates.

Picture the Math Anxiety Process

As you can see, unpleasant encounters with math in formative years can be ruinous to subsequent learning. Students who were made to feel bad about math become wary and prejudiced against it. They mistrust their own abilities. New experiences in math, seen in light of the old, are tarnished by their troubled past, which only accentuates and reinforces long-entrenched negativity. Bad feelings persist. This impairs prospects for learning new material and generates anxiety and self-doubt. They say negative things to themselves, such as "I'm stupid," "I'll never be able to do math," "I'll fail," and "Why do I need to know math anyway?" Soon a continuous flood of negative talk about math ensues; before long, anxiety, overwhelming fears of failing or looking stupid, and panic set in. Physically, these people may experience nausea, perspire profusely, develop a headache or tight muscles, or exhibit a number of other physical symptoms. Mentally, they become confused or disorganized, make lots of careless errors, forget formulas they knew, can't think clearly, or blank out entirely. The end result: poor math performance, avoidance of math, "choking under pressure," and failure. All these negative results lead back to more negative thinking, and the cycle continues. I have diagramed this process in Figure 2-1.

When you deal with math anxiety, you must pay scrupulous attention to what you really think and feel. Negative self-talk seems to be an especially critical habit, which keeps perpetually alive the negative experiences of your past and fosters an ever-increasing anxiety level. Let's begin by looking at your negative dialogue straight on. In the following exercises, you are encouraged to keep a math journal. Writing a journal is an effective means of becoming aware of all your inner mind chatter surrounding math. I encourage you to heighten your awareness of what you say to yourself when you do math. What are the situations and subsequent self-statements that trigger your anxiety?

FIGURE 2-1 THE MATH ANXIETY CYCLE

Stage 1: Contributing Factors

Past negative math experiences; Embarrassments
Negative life experiences associated with learning math
Social pressures and expectations
Desires to be perfect; High demands for correctness
Poor teaching methods; Timed tests; Lack competencies
Negative games people play
Cultural myths; Believing math is innate and not learned
Gender stereotyping and socialization
Negative stereotype threat when asked to perform
Dyscalculia

Stage 2: Negative Thinking and Self-Talk

Negative thoughts about math
Negative thoughts about one's own ability to do math
Preoccupation with disliking math, self-doubts and worry

Stage 3: Anxiety and Bad Feelings

Fears of failure, rejection, or self-exposure
Fears of looking stupid or of being judged or criticized
Feeling incompetent, helpless, worthless, or vulnerable
Feeling nervous and panicky
Poor self-confidence and/or low self-esteem

Stage 4: Physical and/or Mental Symptoms

Mental Symptoms
Negatively affects working
  memory capacity
Confused, disorganized
Can't think clearly,
  incoherent thinking
Makes many careless errors
Forgets known formulas
Blanks out
Easily distracted, can't
  concentrate

Physical Symptom
Muscle tension increases
Headache, shoulder or
  neck aches
Perspiration increases, cold
  sweat
Digestive problems,
  queasiness
Dry mouth, clammy hands
Increased or irregular
  heartbeat
Shortness of breath, feeling
  faint
Shakiness

Stage 5: Math Avoidance or Lack of Math Success

"Choking under pressure"
Poor math performance
Poor progress
Avoidance of math
Failure

Leads back to Stage 2
(and so the cycle continues)

EXERCISE 2-3 **Math Journal Writing**

Starting *today,* keep a journal about your everyday experiences using math. Use a format similar to the chart entitled "My Math Journal," or write in an essay format. Include in your journal the responses to all the questions I have posed. Keep this journal for one or two months. Make at least two entries per week.

Column one of the journal asks, "What situation required me to use math today?" Even if you are not taking a math course currently, there are many daily life situations in which math is needed. Focus on these for your journal writing.

Next, respond to the statement in column two, "What I said to myself and how I felt during and after the situation." Listen carefully to your inner voice and note all your mind talk. Did you say negative statements to yourself about math or about your ability to do math? Then take a look at your feelings. Did you notice any physical symptoms such as sweaty palms or tense muscles in your neck or shoulders? Did you feel dumb or stupid, or confident and assured? Did you blank out, feel panicky, overwhelmed? Record all feelings and sensations.

Column three asks you to review "What I've learned about myself." Objectively look at what you learned from this current math lesson. Did you learn you are too hard on yourself, that you are saying pretty nasty things to yourself about math and your ability to do math? Does one particular thought tend to repeat itself and make you feel defeated? Did you notice you feel unsure even when you can do math problems correctly? Do you give up too readily as soon as the problem becomes a little difficult? Do you know how to do the math, but does the anxiety make you uncomfortable? Write down everything you learn from observing your current math experiences.

The last column asks you to assert "What I'm going to do about it." Now is the time to make a commitment and take some positive actions. Ask yourself: "What can I do to feel better about math? How can I increase my chances of success?" Describe all the steps you are willing to take.

As you continue to keep this journal in the next few months, and progress through this book, more ideas will come to you regarding the most relevant steps you need to take. Jot them down; capture all your ideas on paper.

MY MATH JOURNAL (Sample Form)			
What situation required me to use math today?	What I said to myself and how I felt during and after the situation	What I've learned about myself	What I'm going to do about it

Here are some sample journal entries written by students in my math anxiety reduction classes:

MY MATH JOURNAL			
What situation required me to use math today?	What I said to myself and how I felt during and after the situation	What I've learned about myself	What I'm going to do about it
I had to pay my tuition and medical bills.	"I really should add these up and check if they're correct, but I don't want to." I just didn't feel like adding. Afterwards I felt bad.	Within me, I like and care about math but I'm still ignoring it.	I'm going to work on changing my attitude.
I had to pay a bill for takeout Chinese food for my co-workers.	"I can't do this; I'll just ask a co-worker to add the bill." I felt embarrassed.	I really have a mental block when it comes to math.	I'm going to work on not taking the easy way out and asking others to do my math.

(continued on next page)

MY MATH JOURNAL			
What situation required me to use math today?	**What I said to myself and how I felt during and after the situation**	**What I've learned about myself**	**What I'm going to do about it**
I had to count my hours and figure out my hourly wage to make sure my paycheck was accurate because there was a payroll error the previous week.	*"This is too overwhelming."* *I felt confused and dumb trying to figure everything out.*	*I learned that I need help to conquer my math fears and my block in math.*	*I will take one step at a time to accomplish my goal of conquering my math fears.*
At work we have to measure dimensions on a scale or by digital calipers. A friend was showing me how he did the measurements.	*"I don't understand what he is telling me. I'll just act like I know what he is saying." I felt so stupid and I was so glad when it was over.*	*I learned that I really need to learn math so I can perform my job well.*	*I will continue to study math and not avoid it.*
There was 10% off all clearance sale items. I had to ask the salesgirl how much the price was going to end up being.	*"I should know how to do this."* *I felt inadequate and disappointed with myself.*	*Math holds me back from reacting quickly because I don't understand it.*	*I really want to relearn my basic math and learn to feel more comfortable with math.*

EXERCISE 2-4   ## Collaborative Learning: Share Insights from Your Math Journal

After you have kept your math journal for a few weeks, discuss your insights with a small group of classmates. Share what you've learned about yourself and, most important, what you are going to do about it.

Summary

In this section of our road map, we have investigated the roots of math anxiety and searched into your math history to see where it all began. We have looked at poor teaching methods, embarrassments, negative life experiences, social pressures and expectations, desires to be perfect, negative math games, math myths, gender stereotyping, and dyscalculia. We also diagramed the math anxiety process and saw how negative inner dialogue perpetuates it. You were encouraged to begin a math journal of your current math experiences to gain important insights into how your thoughts and feelings affect your math success. Throughout the chapter, my strong message to you has been: Your past negative math experiences need not continue to burden you. So I urge you to throw off the shackles of the past and, starting today, say to yourself, every day, several times each day, "I can and will succeed in math."

Learn to Manage Anxiety and Improve Working Memory

Imagine, for a moment, a virtuoso concert violinist tuning her masterly crafted violin. When the strings are too loose, the violin cannot be played. The violin sounds flat, sour, lifeless. The tension is too weak, and no matter how well trained the violinist, the violin cannot play beautiful, harmonious music. On the other hand, if the strings are strung too tightly, the resulting sound is a screeching noise, grating and uncomfortable to the ear, and far from pleasing. Just the right amount of tension is needed for the violin to sound its best. And at times during a performance, the violin has to be adjusted and readjusted to maintain this delicate balance. Only through special attention and sensitivity to the fine-tuning of the violin can the violinist be assured of continuous success. The same is true in learning to manage anxiety.

We must learn how to finely tune our anxiety level, for anxiety can be our enemy or our friend. With too little anxiety, we feel lazy and unmotivated. We don't push ourselves to perform or achieve. Our memory seems somewhat dull. We don't think as clearly as we could. On the other hand, when we experience too much anxiety, we feel out of control and devastated. We become tense, out of balance, panicky. Our thinking appears confused and disorganized. We feel insecure, inadequate. Our minds often go blank, and we might experience physical symptoms of muscle tightness, diarrhea, shortness of breath, or vomiting. Images of doom and disaster loom over us.

However, as with the finely tuned violin, when we manifest just the right amount of tension, neither too little nor too much, we think clearly: Our memory is sharp, our perceptions are accurate, our judgments are good. Maintaining the proper amount of anxiety helps us perform at our best. Thus, the goal of anxiety management is not to alleviate all the anxiety you may experience but to help you learn how to manage and finely tune the anxiety you have.

In this portion of our road map, we examine how anxiety affects your memory, and we explore ways to manage anxiety so you can function at your optimal level.

Anxiety Can Affect Your Long-Term and Working Memory Systems

Two memory systems, your long-term memory (LTM) and your working memory (WM), play a crucial role in math learning, mastery, and problem solving. Anxiety can negatively influence both these systems and handicap your ability to solve even the simplest of math problems.

Your LTM has a huge, nearly limitless capacity to store information for long time periods. Its storage bank holds within it all the mathematical facts, principles, formulas, equations, and calculating procedures you have learned while studying math. You can think of the LTM as being analogous to a hard drive on a computer. Your hard drive stores all the information you decide to save so it will be available when you want to access it later.

Students with the best long-term memories figure out ways to relate, categorize, and organize newly learned material with information already in their memory bank so they can retrieve and update the information when necessary. Having ready access to your LTM storage bank and being able to retrieve information easily are essential for success in school.

The LTM system is marvelous, but anxiety can have a devastating effect on it. When students get too anxious, access to the LTM storage bank becomes difficult, if not impossible. It is as if the doors to the memory bank get stuck closed. Then even the most basic formulas or facts aren't available when needed. It's like having your computer freeze up. Your computer hard drive still has all the information stored on it, but you can't get to it until you figure out how to unfreeze your computer. Reducing your anxiety through regular practice of anxiety reduction exercises, such as those discussed in this chapter, will allow the doors of your LTM storage bank to remain open so you can retrieve the needed information.

The second important memory system needed when doing math is your working memory (WM). Your WM is like your own personal processing plant. It is responsible for integrating new information with old memories and processing this newly integrated information for current use in problem solving, reasoning, comprehension, and decision making.

WM allows you to hold information in an active state, long enough so you can utilize the information or process it for long-term storage. WM's storage capacity is limited and only temporary. Cognitive psychologists now believe that individuals can concurrently hold four separate items in their WM or as much as seven items if mnemonic strategies (memory aids) are used (Cowan, 2008; Saults & Cowan, 2007).

Using a computer analogy, you can think of your WM as the RAM in your computer, where you input and work with new information,

**FIGURE 3-1   WORKING MEMORY'S SIMULTANEOUS PROCESSES
NEEDED IN MATH**

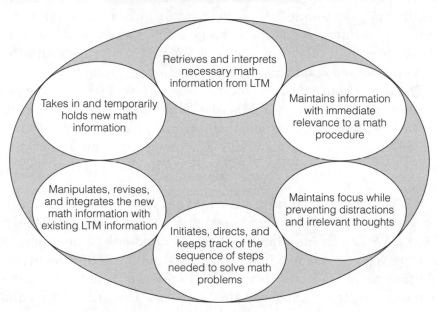

process it, and then save it onto your hard drive or update existing files already on your hard drive.

When working out a math problem, your WM is very busy (see Figure 3-1). It takes in and temporarily stores the new data and instructions for doing the necessary mathematical operations in the current problem, while it is simultaneously retrieving information from LTM, interpreting whether the information is appropriate and relevant for the task, performing the necessary mathematical operations, and retaining results gathered from each procedural step. WM helps you initiate, direct, and keep track of the sequence of steps or procedures you must take. All the while, your WM must help you stay focused and not be distracted by irrelevant thoughts or events in your surroundings.

As you can see, your WM plays a very significant role when you solve a math problem. For WM to function at its best, it needs your conscious effort and your complete, undivided attention. If you are highly anxious or distracted with fears, worries, or negative thinking, your WM capacity will be compromised, and it will be unable to function properly or process the current task. Since your WM is primarily able to hold only four items at a time, if one of those items is taken up with worrying, you would be now working with only three items, or at only 75% capacity. If additional negative self-talk is distracting you, you are now down to two available items, or 50% capacity and so on. Your math learning and problem solving will eventually come to a standstill or barely creep along.

Mark Ashcraft and Elizabeth Kirk (2001) found that individuals experiencing high levels of math anxiety temporarily did not have available to them the full capacity of their WM while solving math problems. This decreased capacity resulted in large increases in reaction time and errors, as well as lowered math performance, when compared with individuals not experiencing this anxiety. The authors suggest that anxiety and intrusive worrying can occupy a significant portion of the WM, preventing it from focusing attention and working on a task. They concluded that "math anxiety disrupts the on-going, task relevant activities of working memory, slowing down performance and degrading accuracy" (p. 236).

Researchers have also found that anxiety is debilitating to students who are most qualified to succeed in math. Beilock and Carr (2005) found that students with the highest working memory capacity and greatest math ability were the ones who were more vulnerable to pressure to perform well. Those students who should have performed best were the ones most apt to "choke under pressure." Their stress and anxiety consumed and overwhelmed their working memory capacity.

Ashcraft and Krause (2007) also confirm that anxiety severely compromises and drains an individual's WM potential. The authors state that "math anxiety seems to influence cognitive processing in a straightforward way—working memory resources are compromised whenever the anxiety is aroused" (p. 247).

This chapter contains exercises to alleviate your anxiety. Regularly practice them, as well as Exercise 5-3 (Challenge Negative Self-Talk) and Exercise 9-8 (Thought-Stoppage Technique), to get a handle on your anxiety and intrusive thoughts and allow you to have full access to your WM capacity.

Fine-Tune Your Anxiety Level

Let's look at what happens when you experience an anxiety reaction. When you begin to feel anxious, many physical changes occur. Your body reacts as if it was in danger and prepares for possible fight or flight. For example, the muscles of your arms and legs often tense, anticipating the need for action; the pupils of your eyes dilate, letting in more light to sharpen your eyesight; your heart rate increases to circulate blood more rapidly to the brain and vital organs; and respiration increases to provide more oxygen to the tissues. Physiologically, hormones are produced that activate the sympathetic portion of the autonomic (involuntary) nervous system. All of these sympathetic reactions would be perfectly suited for survival when dealing with real-life emergencies.

Working out a math problem is obviously not a life-threatening event, but you may be approaching it as if it was. You may begin focusing more and more on fearful, negative self-talk and on the sympathetic bodily sensations. This only serves to increase the anxiety and tension in your body. In a later chapter, we will deal with ways to change the negative self-talk and negative attitudes toward math that feed into this anxiety process and continuously threaten to make it overpowering.

As your anxiety reaction intensifies, the changes in the rate and pattern of breathing tend to be more pronounced. Instead of breathing from the lower lungs, you begin to breathe much more rapidly and shallowly from the upper lungs. You may not even be aware of this more accelerated breathing. As you breathe more quickly, you expel carbon dioxide too rapidly. This produces hyperventilation, a condition resulting in the uncomfortable bodily sensations math-anxious students experience. Some symptoms include confusion, inability to concentrate, shaking, fatigue, muscle spasms or pain, difficulty swallowing, tightness in the throat, choking sensations, shortness of breath, dizziness, irregular heart rate, numbness or tingling of the extremities, and lightheadedness.

As you become more and more aware of these symptoms, they seem to increase. This process, if not stopped in its tracks, can lead to a total panic reaction in which you may completely avoid learning or doing math, freeze up, or actually get sick when approaching math. The following is the most important principle you need to know for managing anxiety:

> By changing the rate and pattern of your breathing (during anxiety-producing situations) to a deep, slow breathing pattern utilizing your lower lungs, you can bring a sense of calm and ease to your body and mind.

Exercises 3-1 and 3-2 will teach you how to calm yourself physiologically and prevent panic from setting in. Instead of automatically breathing rapidly and shallowly with your upper lungs, you learn to breathe gently, slowly, and fully from your lower lungs. These breathing exercises will produce the following calming effects:

decreased heart rate
decreased oxygen consumption
decreased muscle tension
decreased blood pressure
slowed breathing
increased sense of calm and peace in the mind
increased sense of relaxation in the body

EXERCISE 3-1   The Calming Breath

This exercise asks you to breathe slowly and regularly, and with every exhalation, to say the word "relax."

Steps
1. With your eyes closed, sitting in a comfortable position, slowly and gently inhale, concentrating on filling the lower part of your lungs and expanding your abdomen.
2. Now, very slowly and easily, exhale . . . saying "relax" as you do so.
3. Continue breathing gently and slowly, filling the lower lungs and then very slowly exhaling, saying "relax," and feeling more and more relaxed with every breath you take. Continue for approximately 10 more minutes.

I recommend you practice this exercise once or twice each day for 5, 10, or 20 minutes each time. You can listen to this exercise on track two of the CD accompanying this workbook.

EXERCISE 3-2 Deep Abdominal Breathing Technique

This exercise asks you to fill your lower, middle, and upper lungs, while you slowly count to four. Then you are asked to hold your breath to the count of two, exhale slowly to the count of four, and rest to the count of two.

This is an extremely powerful technique during times of panic or anxiety. It should be used for no longer than 5 minutes at a time. However, if you wish to continue longer, you should switch to the Calming Breath technique described in Exercise 3-1.

Steps

1. With this method, it is important to breathe from deep within your abdomen. You may wish to put your hand on your abdomen to feel if you are truly breathing at this deep level. Have your eyes closed and sit in a comfortable position. Now, while slowly counting to four, take a long, deep breath, first filling your lower lungs, feeling your abdomen expand, then filling your middle and your upper lungs. Now hold this breath for the count of two.

2. Now slowly exhale to the count of four while relaxing the muscles of your face, jaws, neck, back, and stomach. And now rest for the count of two.

3. Now begin again. *Inhale* slowly to the count of four. *Hold* to the count of two. *Exhale* to the count of four. *Rest* for the count of two. Continue this exercise for a total of 5 minutes.

Variation 1: The Extended Exhalation

One variation of the Deep Abdominal Breathing technique is to take a slow, deep breath as you count to five, feeling your abdomen rise. Then, very slowly, exhale completely as you count to eight. Continue this breathing pattern for approximately 5 to 6 minutes. If you wish to practice longer, at this point, you may go back to the Calming Breath technique described in Exercise 3-1. You can listen to this variation of Deep Abdominal Breathing on track three of the CD accompanying this workbook.

Variation 2: The Gentle Rhythm of Your Breathing

In this technique, you observe the rhythm of your breathing. Allow your consciousness to become totally aware of your breathing. Notice the temperature of the air, how it is cool when it enters and warm when you exhale. Observe how the air flows into your lungs and then how your abdomen rises and falls with every breath in and out. Continue to focus on this gentle rhythm. Allow it to quiet your thoughts and bring you a sense of calm, peace, and joy. Practice this for about 10 to 15 minutes each day. This is a wonderful technique to relax when you feel yourself tensing up while doing math.

Variation 3: The Extended Sigh Technique

Practice this exercise in a quiet, private place. You may do it standing, sitting, or lying down. Begin by taking a deep, comfortable breath, filling your lungs completely. Then, while exhaling slowly, allow yourself to let out a deep, resounding sigh. Let the sigh extend as long as you can. Enjoy the intense feeling of relief this produces. Practice this deep, extended sigh 10 to 12 times each time you do it. Use it whenever you are feeling a sense of frustration or high stress.

Be sure to practice at least one of the techniques described in Exercises 3-1 and 3-2 a few times a day, every day, for several weeks. These breathing techniques are particularly beneficial when used before entering a math situation in which you are very anxious. By practicing daily, you will become more comfortable and familiar with them so they can be immediately available to you during times of high academic stress.

Our hectic life styles often keep our stress levels so continuously activated that our minds and bodies stay aroused at extremely high levels for days, maybe weeks. Through practicing relaxation exercises on a regular basis, you will be the master of your anxiety. You will be able to quiet your mind and body and restore homeostasis, your body's internal equilibrium.

Studies show the systematic use of relaxation techniques on a daily basis for 10 to 20 minutes produces many significant benefits related to academic performance:

    improved long-term and working memory
    increased measured intelligence
    improved perceptual awareness
    relief from insomnia
    decreased anxiety, irritability, and depression
    increased emotional stability
    improved self-esteem
    greater capacity for reaching one's potential
    improved academic performance in both high school and college

If you are experiencing a great deal of anxiety over math, you will need to create your own special approach. Do it now. If you get a handle on anxiety's earliest signs, it won't get the best of you and grow to panic-like proportions. There is no magical formula to prevent anxiety or panic in math. It will take practice, determination, a positive belief in yourself, and a commitment to reach your math success goal.

Summary

Managing anxiety is like tuning the strings of a violin. When the tension is perfectly adjusted, the resulting music is sweet and melodic. When you learn to fine-tune your anxiety level to allow just the right amount—not too much, not too little—you, too, will perform like a virtuoso.

In this section of our road map, we discussed the important roles that long-term and working memory have when we learn math and how high anxiety negatively affects these memory systems. We also probed the physiological changes you experience during an anxiety reaction. We explored methods for physically calming your mind and body. I encourage you to practice several different breathing techniques. When you are anxious, you must change from a shallow, rapid breathing pattern to a deep, slow breathing pattern using your lower lungs.

CHAPTER
4

Overcome
Internal Barriers

A psychology professor once said to me, "I've noticed that how my students deal with math is how they deal with life's problems." As I pondered what he said, I recognized the truth of his comment.

How do you handle problems when they emerge? Use Exercise 4-1 to see how you deal with life's obstacles. You'll find it most valuable if you're completely honest with yourself and if you practice it before reading further. Have someone read the exercise to you very slowly and calmly, or perhaps you can read it into a voice recorder and later play it back. (This exercise is an adaptation of a brief activity originally described by Susanne Culler in *The Resource Manual for Counselors/Math Instructors,* The Institute for the Study of Anxiety in Learning, Washington, D.C., 1980.)

EXERCISE 4-1 The Wall Fantasy

Arrange to be in a quiet, dark room where you will not be disturbed for approximately 15 minutes. Sit or lie comfortably while you listen to the following fantasy (each set of three dots indicates a few moments' pause):

Begin by getting as comfortable as possible. Settle back, allow yourself to sink down inside, and gently close your eyes. It is important when you do fantasy work to be in a relaxed and meditative state.

Let's begin by focusing on your breathing for a few moments. Take a deep and comfortable breath, filling your lungs completely . . . hold it a moment . . . and then . . . very, very slowly, let it out . . . slowly . . . feeling a wave of relaxation going from the top of your head all the way down to your toes . . . Good! Now take another deep and comfortable breath, filling your lungs completely . . . hold it a moment . . . and then . . . once again let it out very, very slowly, feeling another wave of relaxation going from the top of your head all the way down to your toes . . . slowly . . . feeling more and more relaxed as you do so.

Continue to breathe slowly, deeply, and regularly. (Pause for 30 seconds.) Feel your body becoming more and more relaxed, feeling heavy and more and more relaxed. Allow all the tension to leave your body. (Pause for 20 seconds.)

Now let's begin our fantasy by imagining that you are walking along a road in the country. It is a beautiful day. The sun is shining warmly, but it is not too warm for comfort. The air is crisp and clear, and there is a gentle breeze that feels delightful as it glides past your cheek. The dirt beneath you feels warm and soothing to your feet. The fields on both sides of the road are lush, with vibrant, green, rolling hills. In the distance, you can see trees and wildflowers of every variety and color. Oh, it is a beautiful day!

The road you are walking on leads up and down gentle hills. Feel yourself walking. Feel the gentle breeze blowing through your hair. Feel your arms swinging as you walk along.

The road makes a gradual curve and then, suddenly, your path is blocked by an enormous WALL. Walk up to the wall and feel its texture. As you look up, you see the wall stretching so high it seems to go into the clear blue sky above. To your right, the wall seems to stretch as far as the eye can see. And to your left, the wall stretches so far that it seems infinite. How do you feel? What do you do now? How do you proceed? (Pause for 2 minutes.)

Now slowly bring yourself back into this room. Bring with you the thoughts and feelings you had during this fantasy.

EXERCISE 4-2 Analyze Your Fantasy

Having completed Exercise 4-1, these are some questions I'd like you to think about:

What did your wall look like?

What did you do when you got to the wall?

If you managed to get to the other side of the wall, how did you accomplish this feat?

What do you think is the purpose of doing this fantasy?

What is the most important thing you've learned about yourself that can help you to tackle obstacles you face, particularly with regard to math?

Examine how you deal with the wall in Exercise 4-1 and see how it resembles the way you treat problems in your life. This will give you some sense of how you are handling math. To overcome life's obstacles, we start with an awareness of what we are doing to stop our own progress. Then we must make a conscious decision to change and prevent these obstacles from standing in our path.

When you got to the wall, what did you do? Did you resolutely turn back? Did you cry, feel stymied, not knowing what to do, and so simply do nothing? Did you look for a hidden way through the wall and find a secret door? Did you find a big tree nearby and use it to climb over the wall? Perhaps you dug your way under the wall or started removing stones from the wall to dig your way through it. Did you blast your way through it with a bulldozer or a large vehicle?

If you turned back and didn't go past the wall, perhaps there were times in life you have turned back from obstacles, not looking for ways to overcome them. Have you simply and calmly accepted each obstacle that arises in life, content that maybe this is not the way to go? If this is true for you, what does this mean for how you face the challenge of math? Perhaps you tend to give up too easily when math becomes a bit rigorous, confusing, or in any way anxiety provoking. Or perhaps you tell yourself math just isn't for you and you should choose a major not requiring math. Or instead, you may have been generally complacent about the absence of math in your life, not taking any steps to change this situation.

One student, Diana, said when she came to the wall, she was so frustrated she sat down and cried. She realized she treated her math classes the same way. As soon as a math class got a little hard, she'd get depressed and feel bad about herself. She'd give up, wouldn't try, and eventually drop the course. She had a long history of treating math like this. She knew now she needed to do something different to change this destructive pattern.

Getting Through the Wall

In working with the Wall Fantasy in my classes, I noticed students who are determined to get to the other side of the wall are more persistent in tackling and overcoming their math problems. Look at what strategy you chose for getting through the wall. Were you slow and methodical? Did you blast your way through? Did you find a hidden opening? Did you jump over it as if you were on springs? Looking at your approach to math, how have you or how can you now apply this method to tackle difficulties in math?

Rafael got through his wall by systematically taking out one brick at a time until he made a big enough opening for his body to slip through. He found this slow, methodical approach worked best for him in math, too. Every day, he worked on ten new math problems and five old ones. He went back to basics, reviewing old materials to make sure he had them down pat. Slowly, he chipped away at math as he had chipped away at the wall. His confidence began to build, and he knew he was establishing a good foundation for himself in math. Soon he was able to tackle any new material presented in class.

Elena blasted through her wall the same way she handled problems in life—full steam ahead! But she never thought of tackling math like this until now. She decided math was not going to prevent her from becoming a civil engineer, a career she always wanted. She started reviewing every algebra book she could find. She took a minimum class load and focused mainly on math. She joined a study group, got herself a good tutor, attended a math study-skills workshop, and began studying algebra 2 to 3 hours a day. She practiced relaxation techniques and positive visualization techniques regularly so she could remain calm and positive about math.

Overcoming math anxiety starts with an awareness of what you may be doing now to stop yourself from succeeding. Are you creating your own internal barriers that impede your progress? Are you handling problems as they appear in your path or do you avoid them? Once you realize what you do to hinder your own progress, you can make a conscious decision to take corrective action.

What if you gave up when you came to the wall, and you noticed you have a tendency to give up too easily in math? Ask yourself: Do you really want to overcome your problems with math anxiety? Are you willing to deal with your obstacles *now*? Once you are aware of how you stop yourself from succeeding, you can consciously decide to change this situation. I would suggest you begin by reexperiencing the Wall Fantasy (Exercise 4-1). Take yourself through this fantasy again, but this time do it differently. Try alternative ways of getting through or past the wall. Experiment until you find the most satis-

fying way to overcome this major obstacle. Remember, this is *your wall;* this is your barrier to your future success. You can do it! You can get through it, over it, under it, or even blast it down, leaving it in rubble.

You may also wish to experiment with other techniques to deal with your internal blocks in math. Journal writing, described in Chapter 2, is an excellent method. Exercise 4-3 is a variation on journal writing, and it can help you confront word problems, a nemesis for many. You could also talk with a counselor or math teacher who may be able to give you ideas on how to proceed.

EXERCISE 4-3 Word Problem Log

Many people fear word problems. Faced with one, internal barriers go up, and walls seem to appear from nowhere. As you read the following word problems and begin to work them out, pay attention to how you feel, what you say to yourself, and how this inner dialogue inhibits your progress. Don't worry if you're not sure how to approach these problems; it's not the solutions but how you feel about being asked to solve them that's important. You may be quite surprised at what you learn about yourself.

Sample Word Problems	My Thoughts and Feelings

1. A business executive left her car in an all-night parking garage for three and a half days. The garage charges either an hourly rate of $1.25 or a daily rate of $24 for whole or partial days. Is she better off paying the hourly or the daily rate?

2. Eileen is 4 years older than Monika, who is 2 years younger than Philip. Philip's twin brother, Mark, is best friends with Joe, who is 4 years older than he. Joe babysits for Tom and his newborn baby sister, Julie. Julie is 14 years younger than Joe. How old is Eileen?

Answers to the sample word problems in Exercise 4-3:

1. The daily rate is $24 times 4 days, totaling $96. This is better than the hourly rate of $1.25 times 84 hours, totaling $105.

2. Eileen is 12 years old.

Math Self-Concept and Math Self-Efficacy

Do you have confidence that you have the ability to learn math and do well in math? Your math self-concept is your self-perception of your own competence to learn and do math. Your math self-concept, whether positive or negative, is not necessarily related to your actual math ability. In other words, you may have the ability to learn and to do well in math, but your perception of yourself may be quite the opposite. As a result, if you have a poor math self-concept, you may not put as much effort into math, or you may have more negative self-talk and/or math anxiety when doing math.

Ashcraft and Kirk (2001) found that highly anxious individuals have negative self-perceptions of their math abilities. In their research, they found a negative correlation between students' self-confidence in doing math and their math anxiety. That is, the less self-confident the students were, the more math anxiety they experienced.

I encourage you to examine your perception of your own math abilities. A poor math self-concept may be an important obstacle you must overcome. Exercise 4-4, Your Math Self-Concept, asks you to examine your perceptions of your ability to do math.

EXERCISE 4-4 Your Math Self-Concept: A Self-Assessment

The following statements are all related to your own self-perception of your ability to succeed in math. Read each sentence and decide whether you perceive it to be true, maybe true, or definitely not true for you. Then add up the points.

	True (2 points)	Maybe True (1 point)	Definitely Not True (0 points)
1. I have good math skills.	_____	_____	_____
2. I can solve practical, everyday math problems.	_____	_____	_____
3. I am competent and capable to tutor those with poorer math skills than I.	_____	_____	_____
4. I have a pretty good math mind.	_____	_____	_____
5. I have good math problem-solving strategies.	_____	_____	_____
6. I am confident that I can learn math.	_____	_____	_____
7. I understand math when I am taught.	_____	_____	_____
8. I know that I am capable of math success.	_____	_____	_____
9. I know that I know my math.	_____	_____	_____
10. I feel confident when I take a math test.	_____	_____	_____
TOTALS:	_____	_____	_____

To find your grand total, add up the total points of all three columns:

MY GRAND TOTAL IS: _____

If your score is 14 or higher, you have a generally positive math self-concept.

If your score is 8 to 13, your math self-concept is a little weak and could use a boost.

If your score is below 8, you have a poor math self-concept, and you need help soon! The exercises described in Chapter 5, Positive Thinking Is a Plus Sign, will help you to improve your math self-concept.

Closely related to your math self-concept is how effective you feel you will be if you put effort into learning and doing math. For example, when given a math problem, do you feel that if you try, you can successfully work it out and come to a correct conclusion? This feeling of how effective you would be is called your *math self-efficacy.*

Albert Bandura (1997) found that self-efficacy is dynamic, and it evolves and improves as an individual gains more and more experience with a task. So, potentially, the more effort you put forth as you learn math, the more effective you will feel. Your math self-efficacy will improve, and you will feel more capable of success.

Studies have shown that the higher a student's math self-efficacy, the more likely he or she will be to take math courses and learn math. On the other hand, students with low math self-efficacy are more likely to give up easily or to avoid math. Students who feel that no matter how much they try, they will fail or do badly, will not even try.

I urge you not to allow negative thoughts or a poor math self-concept to get in your way. Don't let them put up barriers for you. Work to improve both your math self-concept and self-efficacy. Put effort into your math and know you can succeed. The more you work on your math, the better you will feel about it and the more successful you will be. To improve your math self-concept and math self-efficacy, follow the suggestions offered in Chapter 5 on positive thinking and in Chapters 8 and 10 on effective math study skills and how to be a successful problem solver.

Summary

In this section of our road map, we examined how we tackle obstacles along our path. We looked at how we deal with life's problems as an analogy for how we deal with math and what we can learn from this experience. We also explored the importance of math self-efficacy and our math self-concept.

Positive Thinking
Is a Plus Sign

Do You Have an Attitude Problem?

Once, when I was a guest speaker in a trigonometry class, a young man named Leonard unexpectedly volunteered a thoughtful and inspiring testimonial:

> I used to do real badly in math throughout high school. I thought math was for the birds. I got so frustrated with it, I couldn't stand it. I began hating it. I felt I just couldn't do it, and I didn't want to even try. I figured I was just a lost cause when it came to doing math. But all that has changed. One day, in a quiet moment, a voice inside me said, "Hey man, you've got an attitude problem." The more I thought about it, the more I realized I was the only one who could change things around. So I made up my mind to look at math like it was doable and important. I saw an image of myself getting good math grades, getting through school, and having a good life. I saw myself using math to solve everyday problems and feeling good about using numbers. And darned if it didn't work! When my attitude about math changed and got better, I started doing great. I like math now, and my work really shows it.

How do you feel about math? Are you intimidated by it? Do you fear it or hate it? Do you sometimes think it's boring or you're wasting your time? Do negative thoughts keep creeping into your consciousness? What do you say when you're having trouble figuring out a problem? Do you say things like: "I'll never be able to do math" "I'm overwhelmed." "Who needs math anyway?" "Why bother?" Do you feel no matter how hard you try, you'll never be able to succeed? Have you ever thought that perhaps your negative attitude toward math diminished your ability to succeed in it?

Educators have known for centuries, for a student to achieve academic success, it takes more than innate ability, competence, or the desire to learn. The key element in this process is having a positive attitude. A positive attitude becomes the catalyst, the supercharger propelling you along the road toward your math goal.

A positive attitude in math is like grade A-1 grease in a well-lubricated engine. The grease enables the engine to run smoothly, with fewer breakdowns or jammed gears, and with less vibration. The engine hums along, working at top speed, doing its job and doing it well. Poorer quality grease would gum up the works, make the engine stick, interfere with its performance, and perhaps totally clog it. Starting today, give yourself A-1 top-rated positive attitudes to lubricate your thinking! The time has come to change your negative attitudes to positive, growth-enhancing, reinforcing ones. Exercises 5-1 through 5-11 will help you do this.

EXERCISE 5-1   Positive Affirmations for Succeeding in Math

An affirmation is a powerful, positive proclamation of something you desire, stated in the present tense. It is a direct, short, simple, active declaration containing no negative words or meanings. It is stated in the "now" as if it is already occurring, even though it might not have happened yet. This is done to reflect an outcome that is desired in the present or as soon as possible.

Read, speak, or visualize positive math affirmations to yourself several times each day. This will help you reprogram and replace negative math self-talk. Some students even record their positive math affirmations in their own voice, with relaxing music in the background, and listen to the recording while commuting to school, running, or going to sleep. This has a very powerful effect!

Some people are bound to say, "This is all well and good, but how do you expect someone to be able to do something he or she really can't do now?"

The point is the skill or capacity to do math, especially if it seems difficult, will come to you much more easily if you begin to believe you're developing it *now*. It's like turning an ocean liner at sea—the ship won't change direction until the helmsman turns the rudder. A positive affirmation is the turning force that sets you off in a new and rewarding direction.

Here is a list of positive affirmations to help you succeed in math. Check the ones you believe would be most helpful on your path to math success. Repeat these statements often to yourself. Remember to substitute these or similar positive affirmations every time negative thoughts about math enter your mind.

1. I'm becoming a good math student.

2. I'm learning more math each day.

3. I'm capable of learning math.

4. I have good abilities in math.

5. I allow myself to relax while I study math.

6. I remember more math each day.

7. I am relaxed, calm, alert, and confident in math.

8. My math improves every day.

9. I understand math if I give myself a chance.

10. I enjoy math more each day.

11. I like math because it's useful in everyday life.

12. Working out math problems is fun.

13. Math is more and more exciting each day.

14. Math is creative.

15. Math is stimulating.

16. My way of doing math is a good one.

17. Math helps me get to where I want to go.

18. Math is my friend.

19. I'm feeling better about math.

20. Math methods help me solve everyday problems.

(Add your own positive affirmations.)

_____ **21.**

_____ **22.**

_____ **23.**

EXERCISE 5-2   Poster Your Wall

Surround yourself with positive thoughts and affirmations. Choose your favorite affirmations from the list in Exercise 5-1 or create your own. Then, with brightly colored markers and letter-size colored paper, put each of these positive math statements on a separate sheet. Write them out in large letters and create a fancy border around each poster. Use decorative typefaces, "WordArt," clip art, and borders to produce fanciful computer posters. Design them as artistically as you can. Have fun with them. Many of my students get a big kick out of this activity. Some of our community college students put flowers on them or smiling faces or fancy designs. Make at least three or four of these posters. Hang them in the place you usually

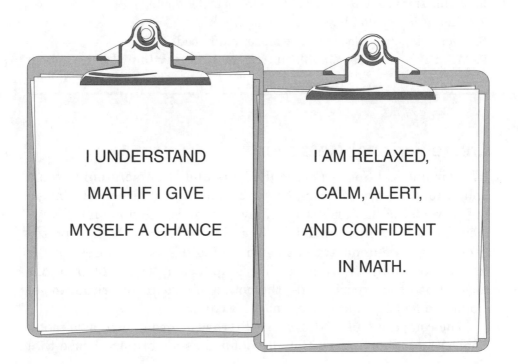

study. Attach them to your refrigerator, on your telephone, on your mirror, in your car, or anywhere that you can see them during the day. Make them visible; they are reminders of your growing positive attitude toward math. My students believe it would have been wonderful if these posters had been prominently displayed, long ago, on the walls in their elementary and middle school classes.

EXERCISE 5-3 **Challenge Negative Self-Talk** (*Optional*)

Every time negative statements about your abilities or about math come to mind, take note of them, and then immediately change or challenge them. Ask yourself these questions:

1. Does this self-talk hurt or help me with math?
2. Does this self-talk affect how I feel about math and my ability to do math?
3. Does it affect my desire or motivation to do math?
4. Is my self-talk based on fact or fabrication?
5. Does this statement distort reality? If so, what is the objective reality?
6. Is this statement 100% true? Is there room for doubt?
7. What evidence do I have to prove it is true?
8. Even if it is true, does it make me, or math, bad?
9. What is a more reasonable statement I can say to feel better about math?

Create Personal Metaphors

In their musical *South Pacific,* Rodgers and Hammerstein wrote a delightful song whose lyrics, on the surface, make no sense: "I'm going to wash that man right out of my hair, and send him on his way." Obviously, the composers did not intend these lyrics to be interpreted literally, yet we understand their meaning precisely. Moreover, the image created by the lyrics conjures up a complex set of emotions most of us have experienced—the notion of a cleansing ritual to get rid of bad feelings and make a positive change.

The composers used the classic literary device known as a metaphor. A metaphor is traditionally defined as a word or phrase that denotes one thing but implies or suggests another. Metaphors are valuable because they help us develop new insights and ways of looking at our human experience. Through their powerful imagery, metaphors inspire and lead us to greater and greater heights and exhilarating emotions. Rodgers and Hammerstein again, in their production *The Sound of Music,* used an uplifting metaphor in the lyrics, "Climb every mountain, ford every stream, follow every rainbow, till you find your dream." It wouldn't be humanly possible to follow these directions verbatim, yet the implied message—to accept challenges, to seek new experiences, to never give up, to live life fully, all in pursuit of our goals—appeals to us at a deeply personal level.

To the extent we act on this message, the metaphor has effectively taught us and changed our ideas. Metaphors help us organize our thoughts and give us a direction for our actions. Not just composers and poets but philosophers and prophets alike have intuitively known and used the power of the metaphor. From Plato's teachings to those of Jesus and Martin Luther King, the metaphor has been a tool to change attitudes and influence behavior. And it still works!

Exercise 5-4 will help you create your own metaphors to alter your negative attitudes and think more positively about math and your ability to do it. In this exercise, I ask you to develop a series of positive thoughts, images, and metaphors for guiding your math success.

First, I want you to list positive adjectives or nouns that come to mind when you think of succeeding in math and overcoming your math fears and anxieties. Some examples include conquest, win, success, enjoyment, innovative, creative, informative, triumphant, victory, important, fun, worthwhile, useful, helpful.

Second, make a list of inspirational images to accentuate these ideas. Examples include guiding light, beacon, buoy, lighthouse, director of a play, completing a puzzle, reaching the finish line, survival, champion.

The third section of this exercise asks you to describe the growth-enhancing emotions you would like to experience. Students who are overcoming math anxiety have listed confidence, relaxation, assuredness, excitement, happiness, thoughtfulness, calmness, courageousness, inner strength.

You are now ready to write your own positive metaphors. Section four asks: What metaphors would inspire you and direct your actions on the road to success? What positive ideas, images, and emotions can you put together to create your own special metaphors? Here are some metaphors that have helped other students on their journeys:

> Reaching for the stars
> Knowing the ropes
> Seeing the light at the end of the tunnel
> Weathering the storm
> Gliding along in cruise control
> Going full-steam ahead
> Bulldozing my way through
> Sailing through with flying colors
> Water rolling off a duck's back
> An award-winning performance
> Going for the gold
> Hitting a home run

EXERCISE 5-4 Create Personal, Positive Metaphors

1. List at least ten positive adjectives or nouns you can associate with success in math.

2. List at least seven inspirational images you can relate to your conquest of math fears and success in math.

3. Describe growth-enhancing emotions you would like to experience as a result of overcoming math anxiety.

4. Now, using the positive ideas, imagery, and emotions you have identified, develop your own special positive metaphors.

5. Once you create your own personal metaphors, use them to inspire and guide you on your path to math success. Picture the images in your mind and carry around with you the positive feelings they produce.

6. Write the metaphors on separate 3″ × 5″ cards and post them in the place you study.

Put Enthusiasm into Your Life

Enthusiasm can make anything we do 1,000 times better. It livens up our lives and our activities. It gives us the "oomph," the energy, and the forward momentum to achieve anything on which we set our sights. When students become enthusiastic about a subject, their motivation increases. They see the subject as their ally and not as their enemy. Things they never noticed before become more meaningful, and learning improves.

Here are seven steps to help you become more enthusiastic about math. Why not give them a try?

1. Delve into Math More Deeply

To become more enthusiastic about anything, you need to increase your exposure to it; you need to explore it and view it from different angles. Start asking more questions, and let your curiosity fly. Search out different math books to further explain a topic you are learning about and to give you a deeper understanding of it. In addition to increasing your enthusiasm, it is a great study-skills strategy.

Begin to pay attention to the usefulness of math in your everyday activities. Look on the Internet for math games and math puzzles. Seek out interesting math facts or curiosities. Look for patterns and relationships; use your imagination. See how math affects every aspect of our lives. Math is amazingly pervasive.

The word *statistics* may have an ominous ring to it, but an astonishing number of average Americans immerse themselves with relish in the batting averages, rebound percentages, and pass-completion ratios of their favorite sports heroes. Probability theory may seem hopelessly esoteric to you; casino owners are counting on this reaction as you part with your money. You may be consumed with the task of equitably dividing up a pizza at a party, yet be blissfully unaware the same theoretical methods apply to apportioning congressional districts.

Like the infinite hexagonal variety of snowflakes, mathematical patterns in nature surround us with beauty and mystery. The spirals on the surface of a pinecone or pineapple, the arrangement of leaves around a plant stem, and the growth spirals of seeds in a sunflower or sections of a chambered nautilus shell all follow the same pattern of elegant mathematical symmetry (known as the Fibonacci sequence). Artists in all cultures have duplicated these patterns in their work for thousands of years.

Fractal geometry describes the shapes of natural objects and boundaries, and modern computer graphics have translated this complex theory into hauntingly beautiful images of imaginary mountains, coastlines, and fantastic shapes.

We are everywhere surrounded and awed by the power of mathematical consistency in an ordered universe. From estimating election results to predicting earthquakes, from designing interconnected telephone systems to analyzing the interconnectedness of the human nervous system, math is, at the core, the one indispensable tool.

Take time to answer these questions for yourself: Why is math so important? What do you believe are the five most essential uses of math in our society? Can you describe at least five ways math is useful in your everyday life? What are three possible ways math can help you in your career choice? Chapter 11 will help you delve into the importance of math in your future.

2. Every Time You Do Math, Liven It Up

Enthusiasm, or its conspicuous absence, affects everything you do. Adding it to your work with math will make you feel better. And even if at first you don't feel like being enthusiastic—*fake it!* It's strange how the mind works. If you start to act enthusiastic, motivated, full of positive drive to do your math, guess what? Your mind starts to believe it, and before you know it, you feel better. So go enthusiastically to math class. Really want to be in class learning the material. Listen and become intrigued, take lots of notes, and ask questions with interest and motivation. Do your homework with gusto. Put vitality into your studying. *Don't* grumble to yourself, "I have to take math." Instead, tell yourself, "I choose to study math—it's a great opportunity." And say it with life! Remember, it's okay to fake it at first because, pretty soon, it will all be true!

3. *Liven Up Your Attitudes about Yourself and Math*

All great successful people know: You are what you think you are. Albert Ellis, in his book *Guide to Rational Living,* tells us our thoughts affect our feelings, and they, in turn, affect our actions. So never underestimate the power of thought.

It is your thought power that directs you in your pursuit of life. If you think you're stupid, if you think you can't do math, if you think you'll never succeed in math—*you're right.* You are what you think you are.

So, beginning today, think positive, enthusiastic thoughts about yourself and about math. Liven up your attitudes. Make them positive and encouraging. Let an optimistic, radiant glow gradually build inside you, a feeling based on the thought, "I can do it!"

4. *Give Yourself Pep Talks*

You are what you think you are—thinking makes it so! So always think positively about yourself and your ability to do math. Never put yourself down or sell yourself short.

Your thoughts, whether positive or negative, grow stronger and more powerful when fueled with constant repetition. You must vaccinate yourself against thinking failure!

What is most important is not how intelligent you are, or what your IQ is, or how much brain power you have. What is really important is how you use what you have. It is not brain power but *thinking power* that counts on your road to math success. The thoughts guiding your intellect are more important than how bright you are. So tell yourself every day, "My positive thoughts and attitudes are my keys to success."

Carry a winning attitude around with you in school, in math class, when doing homework, when studying math. Practice positive pep talks at home, in the car, in bed, in front of the mirror. Discover all the reasons you can to figure out math problems and why it's fun to solve them. Explore how you can be a winner in math. Use your brain power to help search out and create new and better ways to understand and succeed in math. Tell yourself at every opportunity: "I can do math. I'm a great student. I'm a winner!"

Design your own Math Winner's ad campaign to use in your daily pep talks. This is a technique for selling math and you to yourself! It is like the copy (the words) in any advertising campaign or commercial used to sell a new product. In the ad, math and you are the promising new product. I'd like you to buy into this wonderful product and discover the joys of being a winner in math!

Carolyn, a sophomore who is overcoming math fears, wrote this pep talk:

> "Now available—a breakthrough in math students— new and improved! ME!!
>
> "You'll find more to like than ever before. Candidly, I'm bright, uncommonly talented, fabulously motivated, and uncompromisingly determined—everything you've come to expect in a proven math success story. But hold on—there's even more. As a further convenience, I can learn math quickly and readily. Simply stated, I know my basic math facts. But that's just part of the story—I'm starting to really understand algebraic equations. Sounds incredible? And that's not all. I know my formulas and I can easily work out homework problems. As if this weren't enough, my memory is awesome, especially when I stay calm and relaxed. But I don't stop there. You'll be glad to know I'm persistent in my desire to learn and comprehend math.
>
> "The result? I AM A MATH WINNER! Absolutely! Certainly! Fantastic, I admit, but true. Seeing is believing! What's more, satisfaction is assured. And I've saved the best news for last—I come with a lifetime guarantee to succeed in math."

Exercise 5-5 shows you how to develop your own special math winner's "promo." Once you have fashioned it, say it to yourself every day, two or three times a day. It has an ironclad guarantee of success!

EXERCISE 5-5	Collaborative Learning:

The Math Winner's Ad Campaign

1. Think of something positive you can state about yourself in each of the following areas. Don't be modest. Identify what you feel good about, what you do well in, what you feel proud of, what might "endorse" you to others.

    My determination to succeed in math _____

    My motivation to do well_____

    My intelligence level _____

    My ability to understand and comprehend_____

    My memory _____

    The math skills I already have _____

    My "stick-to-it-tiveness" _____

    My desire to learn_____

    My_____

    My_____

    My_____

2. Now write the ad copy to be used in your campaign for selling yourself to you! Talk directly to yourself. Include the positive selling points you identified in part one of this exercise. Be direct. Focus only on the positives.

3. Read your new ad aloud to a small group of classmates.

4. Make at least two copies of this ad campaign. Post one in a place where you can see it often. Carry the other around with you so you can read it frequently.

5. Practice saying your ad aloud, each day, in front of a mirror.

6. Read your ad campaign silently several times every day and before going to bed.

7. Read it before working on math or prior to attending math class. It will boost your spirits and your confidence.

Here is an ad copy written by Raghib, a sophomore at my school.

Raghib, you're doing great. You're a bright guy who works hard at everything you do. You are motivated to succeed and I know you will. You are disciplined and persistent in your pursuit to learn math. Each day you're comprehending more and more math. You are building a strong math foundation using the best math study skills. You have a good memory and fabulous "stick-to-it-tiveness." If anyone is going for the GOLD in math, it is you. Raghib, I'm on your side. You are becoming a math winner!

5. Use the Language of a Math Winner

Whenever you talk about math and your ability to do math, use positive, encouraging terms. Be optimistic in everything you say and do. If someone asks you how you like math and you say things like, "Math is my worst subject," or "It's a constant struggle," or "Don't ask," you only make yourself feel even worse. So instead of responding negatively, say, "I'm starting to really like math," or "I feel great about it," or "My math ability is getting better each day."

Every chance you have, say good things about math. Never put yourself or math down. When you begin to use the language of a math winner, you will be surprised at how rapidly you'll feel better. Tell yourself and the world how great you feel about math, what a worthwhile subject it is, how useful you find math, how it's helping you reach your career goal, and what fun it can be. By using the confident, assured language of a winner, soon you will feel enthusiastic about math and about your potential to succeed in math.

EXERCISE 5-6 **Develop Your Personal Winning Language (*Optional*)**

List five positive statements you could say about math if someone asks you, "How do you like math?" Be sure to use words, phrases, or metaphors that signal hope, promise, joy, cheerfulness, satisfaction, lightheartedness, accomplishment, and victory.

1. _____

2. _____

3. _____

4. _____

5. _____

Use these statements whenever talking about math to others, and say them with enthusiasm!

Here are some ideas of winning language you can use:

> I think math is great.
> I'm liking math more each day.
> Solving math problems gives me a great sense of
>   accomplishment.
> I'm enjoying my math class.
> Math is becoming one of the joys of my life.
> I really love challenging my mind with math problems.
> I'm winning at math and I love it.

---

6. Deal with "Math Downers": Don't Let Negative Thinkers Pull You Down

Don, a sophomore, was surrounded by friends who discouraged him from taking any more than the minimum required math. When he decided to take an advanced math course, they began to tease him mercilessly: "You're a glutton for punishment"; "You're going to become one of those nerds, like the ones we avoided in high school"; "Here comes Mr. Brains"; and "Why struggle, when there are so many more fun courses around?" Don felt torn between studying math and keeping up his relationship with his "math downer" friends. The comments from his friends eventually became toxic to his enthusiasm. He questioned why he needed math anyway. He skipped doing homework to spend more time with his friends, and pretty soon, he stopped coming to class. Don felt like a failure and blamed it on math.

Math downers are everywhere. They have a knack for sabotaging the advancement of those who like math. You must protect yourself from people who put math down and who feed into your own uncertainties and insecurities about doing math. These people are deadly to your progress. They can destroy your plan to successfully accomplish your math goal. They can throw you off track and prevent your journey from reaching a successful conclusion.

So be very cautious. Avoid people who *speak* negatively about math. If you must be with these people, don't talk with them about math. Change the subject. Tell them you'd rather not discuss it. And make it a rule: Never accept advice from people who are down on math. Take anything they say with a grain of salt. Use their negative opinions only as a challenge to prove they are wrong. Don't let them pull you down. Don't let them dampen your enthusiasm.

EXERCISE 5-7 Caution: Beware of Negative Math Thinkers

Who are the people who bring you down or discourage you in math? Identify not only those people who say negative things about math, but also those significant people in your life who just don't seem supportive or interested in your math success. It may not only be the negative words they use or the lack of supportive words, but their body language or tone of voice may also communicate they are math downers. In the space that follows, identify those people you should be cautious of when it comes to math. You may use only their initials if you feel more comfortable. Then, next to the names, jot down effective strategies to handle these people. For example, if you have a parent or a roommate who is discouraging you from taking calculus and has never valued math, you can get support in other ways: Form a math study group on campus, get tutoring at the school's Learning Resource Center, do your homework with a friend in the library, but most important, avoid discussing math at home. What are other productive ways to deal with these math downers?

7. *Build a Support Group of Positive Math Thinkers*

Whenever possible, surround yourself with positive math thinkers, people who like math and are succeeding in it. Attach yourself to these people and let them become part of your math support group. Become their study partners and ask them for advice on homework or points you missed in class. I encourage you to broaden your math support group. A support group is defined as a group of people who support you in the direction you want to go. This will make a world of difference in increasing your enthusiasm for studying math.

Moreover, don't forget you may have an incredible resource already at your fingertips. If you are now in an instructional program, the teachers, counselors, advisers, and math department personnel at your institution are there to assist you, to help you solve your problems, and to further your education. Don't hesitate to use their services. Even if there is, as yet, no formal course or workshop for dealing with math anxiety, your mentors may have a wealth of experience with it and will certainly do everything within their power to further your progress. Make them an active and involved part of your math support group.

EXERCISE 5-8 Identify Your Math Support Group

In the following spaces, identify all those people who are positive math thinkers in your life. List those who you are certain enjoy or love math and think it is fun and exciting. List those people who think it's great that you are working on overcoming your math fears and anxieties. Write down the names of people who are encouraging your progress on the road to success in math. Even if at first you can't think of many people, look around, reach out to others, search for at least ten people who could be part of your positive math thinkers support group.

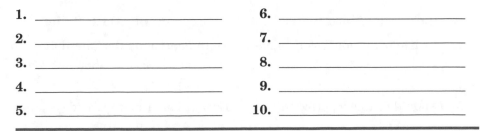

1. _____   6. _____

2. _____   7. _____

3. _____   8. _____

4. _____   9. _____

5. _____   10. _____

Believe You Will Succeed

Whether you believe you can do something or believe you can't—you're right! —Henry Ford

All of our experiences in life are viewed through our own belief-filtering system. If you believe in something, whether it is good or bad, it becomes true for you, and in effect, it becomes a reality. It is like a self-fulfilling prophecy. Once we believe we can do something, we start behaving in ways to make it happen. Our beliefs are like magnets. Once you believe something is possible, your mind seeks ways and means to assure it will definitely happen.

The belief in success is the one great driving force behind all successful students. Believe you will succeed in reaching your math goal and you will! Having a positive belief creates the energy, the momentum, and the means needed to accomplish your goal. It fills you with vitality and vigor to charge ahead and to creatively deal with any obstacle that enters your path.

Beliefs can be very empowering in our lives. They tap into the richest resources deep within us. It is belief that activates your mind to find constructive ways and alternatives to reach your goals. Develop a positive belief system about math and success is sure to follow. Negative beliefs stop you in your path.

EXERCISE 5-9   Rewrite Disempowering Math Beliefs

In the left column, I've listed some common disempowering beliefs that often result in negative feelings and attitudes about math. Add your own disempowering beliefs to this list. In the right column, counter each of the beliefs with a reasonable positive one. I've provided some examples. Use the questions given in Exercise 5-3 to check for reasonableness.

Disempowering Math Belief	**Reasonable Math Belief**
1. Math should come easily to me.	Mathematicians work hard at doing math, so why should it come easily to me?
2. I should do math perfectly.	Even Albert Einstein made computation errors in math.

Disempowering Math Belief	Reasonable Math Belief
3. I should be thoroughly competent in math.	Competence comes through persistence, patience, steadfastness, and diligence. They all take time.
4. There's a right way to do math.	There are lots of okay ways to do anything, including math.
5. I'm dumb when it comes to math.	I'm bright and resourceful and can learn anything I choose to.
6. Math is only for scientists.	Everybody uses math—*everybody!*
7. No one in my family ever succeeded in math, so why should I?	I'm an intelligent and capable person; I've succeeded in a lot of things in my life; why not math?
8. Algebra is not useful in my life.	
9. Math is only for geniuses.	

(Add your own disempowering beliefs)

10.

11.

12.

EXERCISE 5-10 Collaborative Learning: Believe in Yourself

Many students can easily describe why they believe they haven't been able to succeed in math. In this exercise, I would like you to change this process. I would like you to identify and list the reasons you believe you *can* and *will* succeed in reaching your goal. After completing the list, share it with a small group of classmates.

I believe I have the ability to learn.

I believe I am a hardworking, motivated student.

I believe I can learn algebra (geometry, calculus, etc.).

_____

_____

_____

Express Legitimate Math Rights

To act on legitimate personal rights is an essential part of learning to be a more assertive person. Similarly, to express your legitimate math rights is important in affirming both your desire and your ability to achieve math success. Sandra Davis at the University of Minnesota adapted an assertive bill of rights approach for the math anxious.

The list of rights in Exercise 5-11 goes further and constitutes a set of guidelines that may be helpful to you in most situations dealing with math. This is only a partial list, and you may want to add to it. I encourage you to act upon these rights but not to follow them blindly. I suggest you evaluate these rights, change them in ways to suit you, and use them in conjunction with good personal judgment.

EXERCISE 5-11 Identify Your Math Rights

Read aloud the following math rights. Do you agree with them? What other rights should be included? Check the rights upon which you are willing to act. Make a commitment, today, to assertively act upon your legitimate math rights!

_____ I have the right to enjoy math.

_____ I have the right to achieve my math goal.

_____ I have the right to ask questions of my math teachers.

_____ I have the right to ask "Why?"

_____ I have the right to say "I don't know" or "I don't understand."

_____ I have the right to seek help in learning math.

_____ I have the right to be listened to and taken seriously when I ask for math help.

_____ I have the right to learn math at my own speed.

_____ I have the right to see myself as a capable individual.

_____ I have the right to make mistakes in math and to learn from those mistakes.

_____ I have the right to protest unfair treatment or criticism when I'm doing math.

_____ I have the right to be treated as a capable human being by those who teach me math.

_____ I have the right to assess my math teacher's ability to teach me.

_____ I have the right to seek out the best math instruction possible.

_____ I have the right to positive self-regard and a positive self-image, irrespective of how I do in math.

_____ I have the right to remain calm and confident when doing math.

_____ I have the right to work toward achieving success in math.

Other math rights include:

_____

_____

_____

_____

Summary

Maintain a positive attitude toward math and your ability to do math—this is the supercharger propelling you along the road to your math success goal. Utilize positive math affirmations, create personal guiding metaphors, write the math winner's ad campaign, use the language of a math winner, build a math support group, and change disempowering math beliefs to more reasonable ones. These are just a few of the powerful strategies offered in this chapter, strategies to encourage the development of positive attitudes, positive thinking, and increased enthusiasm toward doing math. I encourage you to tell yourself at every opportunity: You can and will succeed in reaching your goal. Tell yourself: "You are a winner and you can do math." Believe you will succeed in reaching your math goal and you will!

Win with "Success in Math Visualizations"

I believe I have discovered the one great, moving, compelling force which makes every man what he becomes in the end.

This, I believe, is the greatest force in the universe. I believe all other causes are secondary to it. It is so powerful that the slightest human effort cannot be put forth until it has done its work; and if it should suddenly be annihilated from the world, all activity would come to a standstill, and humanity would become a mass of automatons moving about in meaningless circles.

This force is not love; it is not religion; it is not virtue; it is not ambition—for none of these could exist an hour without it. . . . It is imagination. —Clarence Budington Kelland (1881–1964)

Your imagination is the creative expression of your mind. It helps produce your present and future reality. Once you understand this marvelous visionary ability within your mind, you can use your visualization skills to construct an environment conducive to achieving math success.

We all visualize or see images in our minds all the time. Take worry, for example: Do you know anyone who doesn't? Of course not. Everyone worries at one time or another. But what do we do when we worry? We imagine the worst thing that could go wrong or that could happen in a particular situation. We put energy into picturing the negative.

Bob worried he wouldn't find a math teacher he liked. He was always afraid that his teacher would go too fast, wouldn't explain things sufficiently, wouldn't understand him, and on and on. His constant worrying exhausted him emotionally and interfered with his ability to enjoy any math class he attended. Bob, with his negative mental images, succeeded in creating his own negative reality. Because of his strong visualization powers, he always saw his teachers in the worst possible light.

In this chapter, we work with your visualization powers. But rather than allow you to put your energy into the negative and visualize the worst things that can happen, I will ask you to visualize the positive and to harness your creative energies to realize your fullest potential.

Much research illustrates the benefits of visualizing or mentally practicing an upcoming event. One such example comes from the sports world. A well-known study (Richardson, 1969) shows how visualization can improve the free-throw scores of basketball players. It demonstrates the positive effects of "symbolic rehearsal" of an athletic activity without any large muscle movements.

Richardson randomly chose three groups of students, all of whom had never practiced visualization techniques. One group of students was asked to practice free throws every day for 20 days. A second group was asked to make free throws on the first day and again on the twentieth, without any practice throwing basketballs in between. The last group was also asked to shoot free throws on the first and the twentieth days. But in addition, this group was instructed to do something quite different. Participants were asked to take 20 minutes each day and visualize themselves sinking baskets. As in real life, when these students saw themselves miss a basket, they were encouraged to work to improve their aim on the next throw.

The results of this study are fascinating. Richardson found that the first group, the students who practiced free throws every day for 20 days, improved their ability by 24%. The second group, who threw baskets on the first and the twentieth days only, as you might expect, showed no improvement at all. The last group, who imagined sinking baskets, improved a whopping 23%! Thus, although they lacked actual physical practice, the last group improved almost as much as the group that practiced for 20 days. Amazing results—and group three might have done even better, if not for one student. This student could visualize the basketball court, but each time he imagined bouncing the ball, the ball stuck to the floor of the court. From this experiment, Richardson concluded it is important to control or precisely structure our imagery. It is best to imagine smooth sailing and the overcoming of any obstacles when using visualization techniques for positive goal achievement.

Program Yourself to Succeed

Emil Coué, a 19th-century French pharmacist, once wrote, "The power of the imagination is greater than that of the will." He said, although you may want something very badly, or "will" it to happen, it won't happen unless you imagine it. You must see it in your mind's eye. You must really imagine it being true for you.

"Programmed Positive Visualization" (PPV) is a technique to help you visualize very clearly whatever changes you want to make in your life and to see your life as you want it to be. It is the deliberate use of the power of your imagination to create your own positive, desired reality. This method can help you consciously change or reprogram your thinking pattern and help you achieve all you desire.

PPV is an incredibly powerful tool when used to improve your math abilities. You might rehearse an upcoming math exam, imagine yourself calm and clear in math class, mentally practice asking your teacher for help outside class, or perhaps see yourself solve equations and difficult problems with ease. You *can* program yourself to succeed in math!

Here are guidelines to set up your own programmed visualization:

1. Decide what you'd like to do or what situation you'd like to improve. State clearly to yourself the math goal you would like to accomplish.

2. Truly want what you imagine. The fewer doubts you have, the greater your visualization power.

3. Some of the relaxation techniques you practiced in Chapter 3 are particularly useful here. When you visualize, allow yourself to be open, positive, and deeply relaxed.

4. Picture yourself doing exactly what you want to do, achieving precisely what you want to achieve, creating just what you want to create.

5. Build a sequence of positive, powerful steps or events in your mind portraying the fulfillment of your goal. Have each step clearly take you in the direction you want to go. Don't imagine difficulties or failures. Picture yourself overcoming all obstacles. Visualize only the progressive movement toward success and achievement.

6. Your visualization will be more effective if you are aware of the sights, sounds, tastes, smells, and the feel of what you visualize. The more complete and lifelike the experience, the better. Some people find they don't actually see images when they visualize, but they can feel, sense, or have the impression or sensation of the images.

7. Choose affirmations to help you feel good and reinforce your progressive movement toward achieving math success. You may choose some positive statements from Chapter 5 or make up new ones. State your affirmations to yourself over and over again during your visualization.

8. Trust and believe what you want will be yours. Push any negative thoughts away. Believe you can and will attain the success you work toward and visualize in your life. Do not discuss your visualization plan with anyone. Avoid exposure to others' doubts or negative thoughts.

9. Once you have designed your visualization, decide on a specific time each day to relax deeply and go through this imagery. Bring yourself into a comfortably relaxed state and repeat your positive affirmations to yourself several times. Next, visualize a gradual progression through the steps leading to your goal, and end with the image of your goal successfully achieved.

10. Each day, no matter where you are—in your car, on your way to class, exercising—visualize your goal and create positive images of achievement in your mind. Always picture the goal as exciting and stimulating. See yourself as successful *now!*

11. Continue this process of relaxation and visualization each day until your math success goal becomes a reality for you. You may then want to repeat this process with a new goal and visualization.

The "Success in Math Visualization" (Exercise 6-1) was designed to help you achieve your math success goal. You can listen to this exercise on track four of the CD accompanying this workbook. Listen to it often.

EXERCISE 6-1 "Success in Math Visualization"

Find a quiet place where you won't be interrupted. Sit comfortably with your eyes closed. Focus on the gentle rhythm of your breathing. Breathe slowly, deeply, and regularly. With every inhalation, feel your abdomen rise, and with every exhalation, feel it fall. Experience all the tightness and tension leaving your body with every breath you exhale. Relax more and more. Go deeper and deeper into a calm and serene state.

Now, with every breath you inhale, say to yourself . . . "I am" . . . and with your exhalation, say "relaxed." As you breathe in, say . . . "I am" . . . and as you breathe out, say "relaxed."

Continue alone for 5 minutes.

Visualize these positive math affirmations as true and say to yourself (pause 5 to 8 seconds between statements):

Deep within my mind, I can see and experience myself relax while doing math.

Deep within me, I can see and feel myself comfortable and confident while in math class.

Every day in every way, my ability to do math is improving.

I enjoy math more and more each day.

I allow myself to remain calm, confident, and comfortable while working out math problems.

Each day, math is easier for me.

I can clearly visualize succeeding in math.

Every day in every way, math is more fun and exciting.

Deep within me, I know I am truly capable and competent to do math.

Deep within my mind, I see and experience the attainment of my math success goal.

Now I would like you to imagine that you are about to do math and that you have your books and notes before you. You have plenty of paper and pencils, and you know you have everything you need. You take a few deep, comfortable breaths and calm yourself before beginning to study and work out math problems. See yourself calm, relaxed, confident. Say to yourself, "I remain calm and relaxed while doing math."

Then visualize yourself concentrating completely on math with your studying progressing well. See yourself understanding your math and having the concepts coming easily and readily to you. Tell yourself, "I can understand math if I give myself a chance."

Picture yourself reading over your math notes. Envision yourself reviewing all the important points and quizzing yourself on the concepts. Tell yourself, "I review all the important concepts over and over again to fix them in my mind."

Imagine concentrating on your work, alert, interested, and enthusiastic. Picture yourself working out math problems, the easy as well as the very difficult and challenging ones. Tell yourself, "Working out math problems is fun."

Imagine yourself, after working for about half an hour, taking a 5-minute study break to refresh your mind and energize your body. Stretch your whole body, reach for the sky, bend down and touch your toes; move around, get some water to drink or an apple to eat. After your break, settle back into your chair, close your eyes for a moment, and take some deep, relaxing breaths. Then visualize yourself resuming study with renewed vigor and confidence. Say to yourself, "Math studying is going well for me."

Now let us conclude your "Success in Math Visualization" by imagining the long-term benefits of confidence and competence in math. Imagine achieving your math success goal. Take the next few minutes to visualize mastery of your goal. Delight in your math success *now!*

How do you feel about yourself and your abilities? How would life be different if you succeeded in reaching your goal? What other benefits would follow as a result of this? Focus on those benefits now! Say to yourself, "Every day in every way, I am working toward achieving my math goal."

Conclude this visualization exercise by saying, "Deep within my mind, I visualize and experience myself achieving my math success goal."

EXERCISE 6-2   Rehearse a Forthcoming Math Situation

Do you get nervous before math class? Does it frighten you to visit your instructor's office and ask for help?

To help you deal with these difficult situations, I suggest you practice this simple technique. Take about 10 minutes to relax and calm yourself down. Do some Deep Abdominal Breathing or the natural Calming Breath described in Chapter 3. Once you feel calm, imagine the difficult situation, and then visualize yourself calmly and peacefully handling it with competence, self-assuredness, and adeptness. See the situation going well for you, and you remain calm and composed throughout. This visualization should be practiced several times in the days before the actual event occurs. It is particularly important to practice it in the half-hour preceding the event.

Build Confidence

Most students who are successful and confident in math or any other subject have learned the one important thing that sustains them: *focus on the positive.*

Successful students have a storehouse of positive images in their memory to readily recall whenever the going gets rough. Past positive experiences are the foundation for future successful ones. If students recall only positive, pleasant experiences related to math, they boost their confidence whenever they need to.

Is your memory bank a powerhouse of positive images?

Here is an exercise to help you deposit only positive images in your memory bank.

EXERCISE 6-3 A Memory Bank of Positive Images

Each night before going to sleep, think of positive, enjoyable math experiences. Don't be hard on yourself. Evoke only good, pleasurable images. Recall even the smallest accomplishments. Perhaps you remember being proud of yourself as a child for learning the times-tables, or when you got the best grade on a math quiz in the seventh grade, or when you did your tax return on your own. Begin to deposit these good math images in your memory. This process will boost your confidence and improve your self-esteem about doing math.

A variation of this exercise is to take a notebook and record all your successes in math. This can help build your sense of accomplishment and self-worth. You need not record major math accomplishments. For example, Dana recorded her correct response when called upon in math class, Marisa wrote she felt successful every time she balanced her checkbook, and Matthew noted he finally understood quadratic equations.

Successful students build their confidence on all the little accomplishments accumulated and stored in their memories. Allow the challenge of math to confront all the strength and power of your memory. Whenever necessary, simply draw on this powerhouse of good math memories. There are no penalties for early withdrawal—only lots and lots of interest!

Anchor Confidence

Wouldn't it be wonderful to always be confident and capable when doing math problems or taking a math class? And if at times you weren't, wouldn't you like the power to just snap your fingers and magically feel self-assured and proficient? Well, you can through anchoring.

Anchoring is a very powerful technique to instantly retrieve past moments when you felt assured, adept, successful, competent, and great about yourself. You can learn to associate these past positive feelings, thoughts, or states of being with any current action or event.

Anchoring is a concept found in the work of Richard Bandler, John Grinder, and other pioneers of *neuro-linguistic programming*. Using anchoring, you can evoke or visualize a past positive success experience and use this experience now to help you feel more positive about math. Anthony Robbins, in his dynamic book, *Unlimited Power* (1997), says anchoring "can create the state you desire in any situation without your having to think about it. When you anchor something effectively enough, it will be there whenever you want it" (p. 274).

Have you ever had the experience of smelling a freshly baked cherry pie, and suddenly, you felt transported in time to childhood when your grandmother had baked you a cherry pie? The good feelings of being with your grandmother, or thoughts of your grandmother's house, may return. In this case, the cherry pie acted as your "anchor" to the past, helping you retrieve the thoughts and feelings of that time in your history. Or perhaps you heard a song you hadn't heard in many years, and it brought on a nostalgic feeling of bygone days. The song acted as your anchor.

Most anchoring occurs without our awareness of the process, but you can consciously use this technique to increase your level of math confidence. Exercise 6-4 will help you do this.

EXERCISE 6-4 **Anchor Confidence in Math**

1. Arrange to be alone in a quiet setting where you will not be disturbed. Sit comfortably with your eyes closed. Breathe slowly and deeply, and gradually calm your body and mind.

2. Now, while in a relaxed state, go back over your life to a time when you felt very successful, extremely confident, unbelievably capable and competent. When you evoke this memory, establish a clear mental image of it. Allow yourself to reexperience this time in your history. What were the sights, the smells, the feel, the sounds, and the tastes? How did you look, act, and respond to others? How did others treat you? What thoughts and feelings did you have? Let yourself physically, mentally, and emotionally "feel" the full impact of this experience of total confidence. Hold your head and body the way you did when you felt this way. Notice your shoulders and the position of your spinal column. Breathe the way you did then. At the peak of this reexperienced state, go immediately to step 3.

   If you can't identify a time when you felt confident and capable, create your own programmed visualization where you see yourself feeling extremely confident and capable. Again, be sure to experience this image fully, on the physical, mental, and emotional levels. At the peak of this experience, go to step 3.

3. At the peak of feeling confident and capable, snap your fingers (or if you prefer, clap your hands, touch your thumb and index finger together, or do something similar) to establish your anchor. As you do this, you anchor the positive state of confidence and capability to the snapping of your fingers.

4. Repeat steps 2 and 3 several times in the next few days with other positive memories or visualizations; each time, get increasingly into the feeling of complete confidence and self-assuredness. At the height of experiencing this strong powerful state of confidence, be sure to anchor the state to your snapping.

5. Soon you'll discover that by just snapping your fingers, you can instantly feel confident and capable. Whenever you are doing math or sitting in math class and you need a confidence boost, snap you fingers! A flood of positive feelings and images will immediately come into your consciousness and help you feel better.

Mary, a premed major, used anchoring to help her feel better about taking math. She had been out of school many years, and when she returned, she found herself very anxious and insecure in her math class. Following the steps in Exercise 6-4, she learned that snapping her fingers instantaneously brought back a stream of good thoughts associated with competence and success. Mary's anchor (snapping her fingers) helped her reexperience the feelings of being the sixth-grade statewide spelling champion, of earning an A on her seventh-grade math final, of receiving an award at high school graduation, and it helped her visualize herself successfully graduating from medical school. Whenever she became fearful or nervous while doing math or when called upon in math class, she'd snap her fingers and a calm sense of success, competence, and assuredness would come to her. Backed by these positive feelings, she could think more clearly, figure out her math more easily, and watch her skillfulness increase.

EXERCISE 6-5 Visualize Metaphors for Success

Assume a comfortable position and close your eyes. Take slow, deep breaths and allow yourself to become increasingly relaxed. In this exercise, I first want you to picture in your mind something about math that makes you feel anxious or uncomfortable. I want you to give life to this image by likening it to an unpleasant but not unfamiliar scene. Here are some examples of images my students have visualized for discomfort and anxiety:

1. Stuck in a deep pit in the ground
2. Feeling claustrophobic in a long, dark tunnel
3. Being in the spotlight in a lineup
4. A wounded animal
5. A tangled, tightly made knot

What images best describe your discomfort?

_____

_____

_____

Now, during your visualization, allow these images to gradually lighten up, fade, soften, and become more relaxing. While you visualize new metaphoric images, repeat reinforcing positive affirmations. For example:

1. While in the deep pit, the sun begins to shine in, and you notice a ladder is tucked away in the corner. See yourself using the ladder to climb out of the pit. Say the affirmation, "I am able to deal effectively with math."
2. The long, dark tunnel suddenly opens up onto a sunny bright seashore with a cool breeze. Repeat the affirmation, "Math is becoming easier and easier for me to understand."
3. The lineup spotlight fades into a soft, glowing lantern, illuminating you under the stars on a romantic evening. Say the affirmation, "I enjoy math more each day."
4. The wounded animal begins to heal and soon is completely recovered and stronger than before. Repeat the affirmation, "I know I am capable, confident, and competent to do my math."
5. The tight knot is loosened and untangled. State the affirmation, "I allow myself to remain relaxed and calm when doing math."

EXERCISE 6-6 My Brilliant (Math) Career

If someone made a home video of your life that you could view later, you would find your memory of the past conflicted with what actually happened.

If you associate a negative feeling with a past experience, realize two things. First, "negative" is a judgment about an experience, but it is not the experience itself. The judgment is made by your conscious mind. Second, realize the only thing that really happened is you formed a perception about an event, and this perception now colors your thinking.

Now that you've developed your own ad campaign to promote your new image as a math winner (Exercise 5-5), go one step further. Become a movie producer and write, direct, produce, and star in the Academy Award-winning movie, *My Brilliant (Math) Career,* which chronicles your life from early successes and triumphs in math to your current state of proficiency and accomplishment. Be sure to dramatize fully the early scenes—in Technicolor, Dolby, and Surround Sound—where you shrugged off a reprimand or made light of an embarrassing moment at school and went on to persevere and prove your math prowess to the entire world (fanfare composed by John Williams)!

Summary

With "Programmed Positive Visualization," you can program yourself to succeed in math. In this section of our road map, we examined strategies to reprogram negative thinking, to rehearse an upcoming event, to build and anchor confidence in math, and to create a memory bank of positive imagery for math success.

Enhance Your Learning Style

As far back as she could remember, Ashley had great difficulty with math. Learning from either the chalkboard or lectures was frustrating. She had trouble concentrating on math for more than 15 minutes at a time. Her mind would wander, and she'd want to do something—anything—but sit in one place. She'd get restless, move around, and want to talk to others, or leave the room. As a child, she was reprimanded for this behavior, and early on, she developed a strong dislike for math.

While working with Ashley, I discovered she needed to be physically involved in the learning process. When she manipulated things with her hands or moved her body while learning, math came easily to her. Hands-on activities, measuring devices, abacuses, beads, seeds, beans, and other physical objects helped her understand and learn math. She needed to walk or pace back and forth as she studied. This helped clear her mind and increase her concentration so she could understand more. I also found she needed to munch on carrots or celery sticks as she studied. Munching kept her energy level and motivation high. So she needed to keep vegetables on hand, cleaned, and ready to eat. Working under these unique study conditions, Ashley was able to bring up her math grade by a grade and a half.

Jon found he studied best by himself, late at night, under a bright light in the quiet of his room. He could study at other times of the day, but night was when he did his best thinking. He noticed he was more creative then, and he could easily solve even the most difficult math problems. Jon also discovered when the room temperature was between 67 and 70 degrees, he could function at his top proficiency level. If the room was much warmer, he became lethargic and sleepy.

Following this study routine, Jon learned his math quickly and more effectively.

What type of learner are you? How do you learn math best? How can you set up your learning environment so you can perform most effectively? This section of our road map will help you enhance your ability to learn math. It is designed to reveal your own personal learning style and the conditions under which you learn best. As you learn math more easily, your math fears will gradually decrease.

Would you like to know what you can do to improve your under-standing and retention of new and difficult math concepts? Would you like to discover what positively affects your powers of concentra-tion? By gaining an awareness of your personal learning style, you'll be able to improve in these areas.

Your learning style has been forged from a unique mixture of personal attributes and preferences, individual background, child-hood experiences, and environmental cues of all kinds. By knowing how these factors influence you, you'll be able to intentionally choose the most effective learning environment and strategies to meet your individual needs.

Learning styles can make a big difference in your life. By un-derstanding and working with your unique learning style, you can greatly enhance your math achievement. You'll study better, feel more excited about learning math, and your test scores will be higher. What's more, you'll feel a greater measure of self-control.

Perceptual Learning Channels

Your perceptual preference may be one of the most significant fac-tors influencing your ability to learn and recall math. Three major perceptual learning channels have been identified: visual, auditory, and kinesthetic/tactile. People who are visual learners learn best if they *see* or *visualize* words and numbers written out. Auditory learn-ers generally learn best if they *hear* math explained to them or if they *say* math to themselves. Kinesthetic/tactile learners need to be involved in the learning process through *touch* or *whole-body move-ment*. You may find one channel is dominant for you and a second one further strengthens your learning.

Professor Rita Dunn, Director of the Center for the Study and Teaching of Learning Styles at St. John's University in New York, has found when students are introduced to new material through the perceptual channel they prefer most, they remember significantly more than when they are taught through their least preferred channel. In addition, if the new material is further reinforced through secondary or tertiary perceptual preferences, they achieve even more.

Here is an exercise designed to help you determine which perceptual channel you prefer most for learning math.

EXERCISE 7-1 Your Perceptual Learning Channels: A Self-Assessment

Carefully read the sentences in each of the following three sections and note if the items apply to you. Give yourself three points if the item usually applies, two points if it sometimes applies, and one point if it rarely applies.

Are You a Visual Learner?

_____ 1. I am more likely to remember math if I write it down.

_____ 2. I prefer to study math in a quiet place.

_____ 3. It's hard for me to understand math when someone explains it without writing it down.

_____ 4. It helps when I can picture working a problem out in my mind.

_____ 5. I enjoy writing down as much as I can in math.

_____ 6. I need to write down all the solutions and formulas to remember them.

_____ 7. When taking a math test, I can often see in my mind the page in my notes or in the text where the explanations or answers are located.

_____ 8. I get easily distracted or have difficulty understanding in math class when there is talking or noise.

_____ 9. Looking at my math teacher when he or she is lecturing helps me to stay focused.

_____ 10. If I'm asked to do a math problem, I have to see it in my mind's eye to understand what is being asked of me.

_____ TOTAL SCORE

Are You a Kinesthetic/Tactile Learner?

_____ **1.** I learn best in math when I just get in and do something with my hands.

_____ **2.** I learn and study math better when I can pace the floor, shift positions a lot, or rock back and forth.

_____ **3.** I learn math best when I can manipulate it, touch it, or use hands-on examples.

_____ **4.** I usually can't verbally explain how I solved a math problem.

_____ **5.** I can't just be shown how to do a problem; I must do it myself so I can learn.

_____ **6.** I've always liked using my fingers and anything else I could manipulate to figure out my math.

_____ **7.** I need to take lots of breaks and move around when I study math.

_____ **8.** I prefer to use my intuition to solve math problems, to feel or sense what's right.

_____ **9.** I enjoy figuring out math games and math puzzles when I learn math.

_____ **10.** I learn math best if I can practice it in real-life experiences.

_____ TOTAL SCORE

Are You an Auditory Learner?

_____ **1.** I learn best from a lecture and worst from the chalkboard or the textbook.

_____ **2.** I hate taking notes; I prefer just to listen to lectures.

_____ **3.** I have difficulty following written solutions on the chalkboard, unless the teacher verbally explains all the steps.

_____ **4.** I can remember more of what is said to me than what I see with my eyes.

_____ **5.** The more people explain math to me, the faster I learn it.

_____ **6.** I don't like reading explanations in my math book; I'd rather have someone explain the new material to me.

_____ **7.** I tire easily when reading math, though my eyes are okay.

_____ **8.** I wish my math teachers would lecture more and write less on the chalkboard.

_____ **9.** I repeat the numbers to myself when mentally working out math problems.

_____ **10.** I can work a math problem out more easily if I talk myself through the problem as I solve it.

_____ TOTAL SCORE

My dominant perceptual learning channel is:

_____

(enter the category with the highest total score)

My secondary perceptual learning channel is:

_____

(enter the category with the second highest total score)

My tertiary perceptual learning channel is:

_____

(enter the category with the third highest total score)

Suggestions for Distinctive Learning Channels

Visual Learners

Are you a strong visual learner? Do you find you must see math problems written on the chalkboard or on paper before you can begin to comprehend what is being asked of you? Would it drive you crazy if you had to listen to a math lecture and you had nothing to write with or if the teacher wrote nothing on the board?

Ted, a construction worker, returned to college after being out of school for almost 10 years. He really wanted to get his college degree, but math was terribly frustrating to him. He could easily copy everything his instructor put on the chalkboard, but he was completely lost with the lecture. Math just didn't make sense to him when he listened to it. Ted learned he is a strong visual learner but a weak auditory learner. Once he understood _how_ he learned, he began using strategies to help him gain the most from his math classes and his studying. Soon Ted's math achievement and his enjoyment of math began to improve. Ted has continued to take math courses and is now doing well in calculus and loving it.

These strategies will help if you are a strong visual learner but are weak in the auditory channel.

1. Always take written notes when someone is explaining math to you.
2. Whenever possible, ask for written instructions.
3. Make your own drawings or diagrams when figuring out word problems.
4. Use flashcards to review all important concepts, formulas, theorems, equations, and explanations.
5. Write as much as you can when you study. Work out lots of problems.
6. In lecture, concentrate on what the instructor is writing on the chalkboard and copy everything down. If you can't get much of what the teacher explains in class, bring a voice recorder. Always reset the counter on the recorder to zero at the beginning of the lecture. At points in the lecture when you don't fully understand what the teacher says, note the counter number on the recorder and put it down in your notes. Later, you can listen carefully to the recording, paying attention to the sections where you jotted down the counter numbers. Write down the information that you missed getting the first time.
7. Use two or more math books. Read how different authors explain the topics you learn. Because this is such a good study-skills technique in general, I will discuss it in further detail in Chapter 8.
8. Visualize in your mind's eye the math concepts you are learning.
9. Whenever possible, use computer programs to illustrate concepts you are learning.
10. Read your textbook assignment and previous class notes before your next class.
11. Use workbooks, supplemental study guides, CDs, videos, or handouts.
12. Map out, chart, or in some way graphically illustrate your classroom and textbook notes.

13. Always write in your textbook. Underline key words. Mark important concepts and use colored pencils to liven them up.
14. Sit near the front of your classroom to avoid visual distractions and to pay closer attention to your instructor.
15. When you review your classroom notes, creatively highlight the important points with colored pencils or markers.

Auditory Learners

Are you primarily an auditory learner? Do you prefer to have someone explain math to you rather than read about it or see it on paper? Do you often have to repeat math problems aloud or in your head before you can figure them out? Do you just hate it when a teacher shows the class how to figure out a math problem on the board but doesn't explain each step aloud while writing it?

The following suggestions will be particularly helpful if you are a strong auditory learner but are weak in the visual area.

1. Sit near the front of the classroom so you can clearly hear your teacher without auditory distractions.
2. You may want to use a voice recorder during lectures and listen to each lecture as soon after class as possible. Listen to it over and over again, when you drive, study, jog, or do your chores.
3. Take part in classroom discussions.
4. Ask lots of questions in class, after class, and in help sessions. Ask for clarification if you don't completely follow an explanation in class.
5. Restate, in your own words, math concepts you are trying to understand.
6. Ask your math teacher to repeat important concepts.
7. Listen carefully to the math lecture. Mentally follow the concepts and then write them down to capture what was said.
8. If you can't get everything the teacher writes on the chalkboard, find a classmate who seems to be a visual learner and is writing down everything from the board. Ask if you could photocopy this person's class notes.

9. When figuring out a difficult homework assignment, you may want to read it aloud into a voice recorder and then listen to it and write it down.
10. Immediately after you read your math textbook assignment, recite aloud what you have just learned.
11. Read your class notes and textbook notes aloud. Whenever possible, say them in your own words into a voice recorder.
12. Talk about math to a study partner or to anyone who will listen. (I know some students who have even explained their assignments to their pets.)
13. Listen for key words in your math lecture. Note if your instructor emphasizes certain points through tone of voice, enunciation of certain words, voice inflections, and so on.
14. Record all the key concepts, formulas, explanations, and theorems on a voice recorder, and listen to them often.

Kinesthetic / Tactile Learners

Was your score on the perceptual learning channel assessment highest in the kinesthetic/tactile area? Do you prefer a real-life experience with math, manipulating it and experimenting with it? Do you find you like to move around when you study, pace the floor, or shift positions a lot?

These strategies will help if you are a kinesthetic/tactile learner:

1. You must use a hands-on approach to learning. Work out as many math problems as possible. Do, do, do. Practice, practice, practice. You'll be amazed at the positive results.
2. Whenever possible, convert what you are learning in math to real-life concrete experiences. If applicable, use measuring cups, measuring vials, toothpicks, seeds, stones, marbles, paper clips, rulers, sticks.
3. If someone shows you how to do a problem, immediately ask if you could work out a similar one to see if you understand how to do it.
4. While studying, try to solve problems several different ways to decide which method feels right to you.

" JOAN'S A KINESTHETIC LEARNER "

5. Many kinesthetic/tactile learners find they must move during the learning process. You may want to walk to and fro while reading your assignment or even while working out problems. Some students like to rock back and forth. Others need to shift positions frequently. The movement increases understanding and comprehension for some highly kinesthetic people.
6. Use computers and workbooks.
7. While you exercise or engage in other types of physical activities, review your math concepts in your mind.
8. Use your fingers and even your toes if this helps when you figure out math problems.

9. Rewrite class notes.
10. Use a calculator to solve problems.
11. If possible, use or build models to help you understand the math concepts you learn.
12. Study math on an exercise bike—preferably one with an attached reading stand allowing you to move your arms and legs.

Strengthen All Areas

I recommend you strengthen and interrelate all your perceptual learning channels. For example, recite aloud what a diagram or chart represents so you both see and hear the information. You may also redraw it or write out an explanation of it so you can be kinesthetically involved. Each channel reinforces the other. This is particularly helpful because you may not have the luxury of being in a course where the instructor teaches to your preferred learning style. I believe that if you strengthen all modes of learning, you'll be better able to process information on both sides of your brain. Then, if you don't recall the information you've learned one way, you may be able to recall it through another. So see the information, say it, hear it, write it down, feel it, visualize it, read it, and manipulate or work with it in as many ways as you can.

A Sequential versus a Global Learning Style

When learning math, it is important to be aware of how you take in, interact with, and process the new information and demands of the situation. If you understand your approach to new material, you can explore ways to adapt to the requirements of the task.

Sequential learners tend to learn in small incremental steps. They follow a step-by-step approach; using little pieces of knowledge, they learn to advance them in a logical, linear order to the next step, then the next step, and so on. This sequential approach helps them solve math problems without necessarily fully understanding the math concepts with which they are working.

If this learning style describes you, you are fortunate because throughout the educational system many teachers and textbook authors use a sequential approach when presenting material. Math textbooks are usually presented in a sequential order, as are many lectures, lesson plans, syllabi, and math curricula.

On the other hand, this approach might not characterize you at all. Perhaps when you take in little bits and pieces of math knowledge, they appear disjointed or unrelated. You don't see or appreciate a step-by-step logical sequence. The information presented seems irrelevant and doesn't make sense. You feel as if you are stumbling around in the dark, progressing slowly or not at all. But you may have had the experience where you suddenly see the whole picture clearly. Everything miraculously comes together like a puzzle. Aha— you get it! In one fell swoop, you finally see it all. All the bits and pieces now make sense. You have found the light switch after trying all the switches in the dark mansion. Now you can see the room in which you are standing, but another surprising thing happens. You can see into the adjacent rooms, and the rooms beyond those, and even into the grounds outside. The light illuminates far more than your immediate surroundings.

What I've just described is the experience that many global learners have had. Once global learners grasp the concepts, they tend to make connections and have a more universal view of what they learn than other types of learners. They can see the interrelation and application of these new concepts to other things they already know. They may be able to solve difficult problems in new and innovative ways once they obtain an understanding of the whole. Explaining how they achieved an answer can be difficult for global learners because of the huge leaps in understanding they are able to suddenly take.

Global learners have the potential to be very creative thinkers not only because they can take large leaps in their understanding, but also because they can apply what they have learned to other areas in their studies, careers, or personal lives. So being a global thinker definitely has its advantages. But it has its disadvantages also.

The problem global learners experience is that it takes them quite a while before they can see how the bits and pieces of information fit into the greater scheme of things. Without grasping the entire picture, the small bits of information seem senseless to them. They may struggle with new concepts, homework, and problem solving.

Since global learners don't learn in the traditional manner, these learners may feel dumb, out of sync with their classmates, and unable to perform well or meet their instructors' expectations. They may become discouraged when they try to grasp the whole picture and can't quite get it, while their classmates appear to function quite well.

If you are one of these global learners, I encourage you not to give up when learning math. Both the sequential and the global approaches to learning are important. As I have said, global learners have the potential to be very creative because they can take large leaps and apply what they have learned to other areas.

Here are some suggestions to help you if you are a strong global learner:

1. Continue to learn new information, even if you don't readily see the whole picture. Don't give up.
2. Keep taking in as many facts and bits of information as you can, until eventually you'll be able to grasp the complete picture. Before you tackle a new chapter in your textbook, skim the whole chapter and get a feel for its overall theme.
3. Be patient with yourself. Know that the big picture will appear. Know that understanding will come.
4. When you learn a sequence of steps in a math operation, ask your teacher or your math tutor for the operation's goal or overview. Ask for help to see the whole picture.
5. Work with your teacher, a tutor, or other classmates to explore how you can apply the new information you learn to other new concepts, to concepts you previously learned, and to ideas in other courses and disciplines. Seek out ways this new information is relevant in your everyday life.
6. Whenever you have a chance, give yourself a large chunk of time to study your math. Extra time gives you more of an opportunity to synthesize, connect, and make sense of the many bits and pieces of information you have learned.
7. Understand that you have a unique ability to put things together in new and different ways, and this can be a great advantage in many careers and jobs.
8. Relate what you learn to other courses, such as chemistry, physics, music, or accounting.

9. Learn to appreciate the sequential learners' approach to problem solving; that is, how each step can build on the next and lead in a logical, linear fashion to a problem's solution. Your global approach need not be static. When you use both the global and the sequential learning approaches, you are more likely to increase your success in math.

Here are some suggestions to help you if you are a strong sequential learner:

1. Be aware that most math textbooks and math lectures follow a logical, linear approach. If the step-by-step linear sequence is not apparent to you in class, ask your teacher or a math tutor to help you see the flow. Ask for help to fill in any missing steps or "in-between" smaller steps that would help you in your progress toward your goal. You can also consult different math books that explain the same topic and may explain the smaller steps that were absent in your lecture.
2. Drill yourself on the basic procedures you learn so they come easily and readily.
3. When you study and do homework, write out the step-by-step procedures for ways to solve problems, and use this approach to outline your class notes.
4. Learn to appreciate the global learners' approach. Work to synthesize everything you learn so you can get the bigger picture of how the new information fits into previously learned concepts and into this course material and related disciplines. This will deepen your understanding.

A Deductive versus an Inductive Learning Style

From the time he was quite young, my son's style, when given a new electronic device, was to start to use it immediately, before reading the directions. This approach rapidly solidified his functional grasp of that machine's operation. Written materials were used only as a last resort ("when all else fails, read the instructions").

Contrast this with the style of the person who will not touch a gadget before mastering all written directions, methodically working through them step by step, and who actually may be intimidated by its mechanical aspects ("if all else fails, plug it in").

In between are mixtures of the two extremes—those who combine a common-sense willingness to follow written instructions with a spirit of adventure and a readiness to derive flashes of insight from a trial-and-error approach.

When it comes to learning and applying mathematics, where do you fit into this broad analogy? Is your personal thinking style to immediately begin to work out and solve problems, allowing the insights you gain by doing to give rise to the broader generalizations and principles that apply to all similar problems (inductive process)? Or do you prefer to assimilate theory and background until you have a clear concept of the principles involved, and only then apply them to individual problems one at a time (deductive process)?

Doers

Are you an inductive "doer"? If so, you may prefer to do math problems before you've understood your assignment or the principles involved. If this characterizes you, I suggest when you are presented with a new math topic, work out lots of problems to get a good feel for that particular concept. Attempt to identify the commonalities among the problems so you can ascertain the unifying principles involved and later apply these principles to other, more difficult problems.

Each time you figure out a new principle, test it out to see if it really works. See how many different ways the new principle you've learned can actually be used in real-life examples. This is an excellent approach to enhance your learning, but at times, even this approach may not be adequate. For example, many doers experience anxiety when faced with problems for which the solutions require new knowledge and not just number crunching, thus underscoring their lack of information. If this happens frequently to you, it would be helpful to go back to basics, recognize the deficiency, and develop the good study-skills habit of reading the assigned material before attempting to do homework problems.

Cogitators

Perhaps you're a deductive "cogitator." If you are, you may prefer to grasp theory and principle before attempting to do any of the assigned problems. This is an excellent approach because it's certainly easier to memorize math rules if you understand them first. But math concepts are often hard to fathom at the onset, and this may cause you to get stymied and experience anxiety or distress. If this happens to you, try easing into problem solving like sliding into a cool stream, gently testing the water. Sometimes, only by working out problems do certain principles begin to make sense. First, work through sample problems in your textbook that show the solutions step by step. Do as many sample problems as you can from your text as well as from other textbooks and college course outline series. From there, you can move on to problems of progressively increasing difficulty, always doing only problems for which the book offers the solution. Keep going back and forth between reading the principles and seeing how the principles are put into action. This process will help you better understand the concepts and principles you're learning. Whenever you feel stymied by a new concept, you need to start to put the principles into practice. Soon you'll feel much more comfortable.

EXERCISE 7-2 Collaborative Learning: Share Your Insights on Learning Styles

After you read about visual, auditory, and kinesthetic/tactile learners, global and sequential learners, and deductive and inductive learners, write your insights in the space provided and share them with a small group of classmates. Describe your learning styles and the strategies you've found most helpful when you learn new math concepts and procedures.

Insights:

_____

_____

_____

_____

Other Major Factors Affecting Your Learning

We are all biological creatures affected by a bewildering assortment of internal and external stimuli. We get hungry and we eat; we get full and sleepy; we are startled by noise or lulled by it; we fidget with energy or drowse in repose. Our attention is diverted by so many distractions it's sometimes hard to imagine getting our brains to focus on learning new material. But learn we must. And if we want to do so most efficiently, we must analyze these influences and either minimize the disturbance they cause or maximize their positive effects.

Time of Day

When is your energy at its highest? Are you an early bird ready to go with eagerness and vigor at 6 or 7 in the morning? Or are you a night owl like Jon, who was mentioned earlier in this chapter? Maybe you feel as if you're in a fog until 10 A.M., and it would be a crime to force you to take an 8 or 9 A.M. math class. Perhaps you become a shining star at noon or in the early afternoon. Or possibly, you are roaring your engines at 7 P.M. and can go full steam ahead until midnight.

Lanny, a student in my math anxiety reduction course, said she became pretty discouraged with math in high school. She was always scheduled to take math first thing in the morning, usually a 7:40 A.M. class. She found she could never concentrate in class and couldn't grasp the concepts. She began to hate math. Now, when she looks back at her high school years, she realizes she was a confirmed night owl, staying up late each night and never really coming to life the next day until around 11 o'clock in the morning. Lanny is still a night owl, but now she makes sure she schedules her math classes in the afternoon or evening, and she never takes morning classes. She studies best after everyone in her family is sleeping soundly.

Each of us has times in the day or evening when we perform at our peak level of proficiency. If you schedule your math classes or math study sessions at these times, you'll find you can concentrate better and learn more. Exercise 7-3 will help you determine the hours of your peak performance.

EXERCISE 7-3 **When Is Your Mental Energy at Its Highest?**

Notice your mental energy level throughout the day for a few days. On the chart below, use one check to indicate the times you feel alert, clear, and can concentrate and study well. Use two checks when you function at your peak: think best, learn more quickly, concentrate deeply, understand clearly.

	Day 1	Day 2	Day 3	Day 4	Day 5
5 AM					
6 AM					
7 AM					
8 AM					
9 AM					
10 AM					
11 AM					
NOON					
1 PM					
2 PM					
3 PM					
4 PM					
5 PM					
6 PM					
7 PM					
8 PM					
9 PM					
10 PM					
11 PM					
MIDNIGHT					
1 AM					
2 AM					

Sound Level

Do you like to study in a very quiet place, free from all distractions? For most people, this increases comprehension and the ability to figure out difficult problems.

Francine preferred silence and was easily disturbed by any sounds, particularly when she was nervous about an upcoming math test. She found wearing small moldable earplugs, which she bought at a local pharmacy, did the trick. She wore them every time she studied at home or in the library. Her concentration and retention levels increased markedly.

Other students find silence especially helpful when they work on demanding or problematic assignments, but they prefer some background music or noise when they do a routine or boring assignment or when they recopy notes.

Still others find, when it's too quiet, they become "hyperaware" of all sounds. Howard, an accounting major, always studies with the television or radio on in the next room, although he doesn't listen to it. If it's too quiet, he hears the refrigerator motor, the ticking of the living room clock, and the sounds of his own heartbeat.

Harriet, a statistics student, found playing soft, inspiring music increased her motivation for studying and the amount of material she covered. Tony, a calculus student, learns best when listening to music of classical baroque composers. The slow tempo of approximately one beat per second calms him and clears his mind.

EXERCISE 7-4 **How Does Sound Affect You?** (*Optional*)

Briefly describe which sound conditions work best for you when you study math.

Lighting

How does lighting affect your ability to study and learn? Do you study best with natural light coming in from the window? Or maybe you like to cuddle under a soft incandescent lamp. Perhaps you function best under white fluorescent bulbs or a bright halogen light.

Some students find bright lights energize them and make them more alert and attentive. When the light is subdued, they lose interest easily and become distracted, apathetic, and drowsy. For other students, bright lighting has the opposite effect; it makes them tense, fidgety, and uncomfortable.

Many people feel more relaxed and concentrate better when they work in diffused natural light or under balanced full spectrum fluorescent lights. Experiment with various lighting conditions and figure out for yourself how lighting affects you.

EXERCISE 7-5 **How Does Lighting Affect You? (*Optional*)**

Briefly describe which lighting conditions work best for you when you study math.

Temperature

Do you prefer to learn in a cool or a moderately warm room? Some students think better when the temperature is cold, whereas others find it intolerable and distressing.

Each of us has our own unique reaction to temperature. What if you're taking math in a warm classroom and you think better in the cold? Why not wear very lightweight clothes to class? Consider sitting near a window or an open door where you might get some cooler air, and be sure not to sit under heating ducts. If the opposite situation occurs and the room is too cold, take extra layers of clothing to class to keep yourself warm. You also might notice at different times of the year or at different times of day, your reaction to environmental temperature varies. If you're extra tired, catching a cold, or disturbed by a personal problem, you might feel colder than usual.

EXERCISE 7-6 **How Does Temperature Affect You?** *(Optional)*

Briefly describe which environmental temperature conditions work best for you when you study math.

Room Design

Do you like to study in a big, soft, comfortable chair or couch? Or maybe you study best at a desk, seated in a nicely cushioned straight-backed chair. Perhaps you like to sprawl out on a carpeted floor with papers and books spread out around you and soft, fluffy pillows to lean on. For some students, this eases their tension and motivates them to study more.

Andrea found if she sat in a comfortable armchair, she soon became sleepy and her eyes closed. She became too relaxed, and she rapidly lost all motivation for studying. To do her best, Andrea arranged to study at a large, uncluttered desk where she could use the entire top surface to spread out her math books, assignments, class notes, scratch papers, and practice worksheets. This arrangement helped her to see everything clearly and to stay in a problem-solving mode. At school, if she sat in a study cubicle in a quiet corner of the library, she stayed right on target. She avoided the large, comfortable library armchairs where she saw some of her classmates snoozing.

When you prepare for tests, take your practice exams in a room arranged much like the testing room. So, although you might prefer to learn in a comfortable setting, at a large desk, or in the middle of your bed, you must be able to transfer this learning to the classroom and testing situation. This transfer will occur more easily if you give yourself practice exams in a room of similar design.

EXERCISE 7-7 **How Does Room Design Affect You? (*Optional*)**

Briefly describe in what setting you learn and concentrate best.

Food Intake

Do you often have the "munchies" when you study and can't concentrate unless you have something in your mouth? Or perhaps you must wait at least an hour or two after you eat before you can concentrate on learning anything.

   You may have noticed certain foods affect your study capabilities more than others. Too much caffeine may make you nervous and jittery, and a large fatty meal may dull your senses and make you feel exhausted and lethargic. A few ounces of protein eaten an hour or two before you study may help increase your alertness and motivation. An ounce or two of carbohydrate eaten before or while you study can calm you and increase your focus. Chapter 9 discusses some of the effects of food on the thinking process.

EXERCISE 7-8 **How Does Food Affect You? (*Optional*)**

1. Do you think better if you snack when you study?
2. What foods seem to affect you positively?
3. What foods negatively affect your ability to study or make you lethargic and tired?
4. After you eat a meal, how long does it take for you to reach your best level of concentration?

In the following space, briefly describe how food affects your ability to study math.

Mingle or Single

Are you a loner when you study or, would you prefer studying with others? Having a "study buddy" can be an effective study tool. A study partner can also alleviate feelings of isolation or loneliness. For ideas on how best to benefit from studying with others, be sure to read the section "Study Buddies" in Chapter 9.

Some students like working in small groups with classmates and rave about their study group. Others have been disappointed with them. Still other students find studying a very private act. They prefer no one to disturb them. They lock themselves away in a library cubicle or in their bedrooms.

I've also counseled students who prefer a combination of being alone and being with others. They like to study by themselves but with the hustle and bustle of activity and movement around them, as long as no one pays attention to them. These students often enjoy studying in a coffee shop or the school cafeteria. Many report they enjoy studying in their school math tutoring lab where they can be alone but where there are others who are also studying math and where tutors are available if needed.

Clothing

Do you prefer to wear loose, comfortable clothing when you learn math, or is this not an issue for you? In class and when taking tests, make sure you wear clothes that are as comfortable as the clothes in which you study. What are your favorite clothes for studying? What clothes make you feel successful and sure of yourself?

Summary

You can greatly enhance your ability to learn math if you remain aware of, and at times also manipulate, the factors that influence your unique learning style. In this section of our road map, we assessed whether you are a visual, an auditory, or a kinesthetic/tactile learner, and I offered you learning strategies specifically adapted to your preferred perceptual learning style. We also explored the sequential and global learning styles, as well as the deductive and inductive learning styles. Then we examined how you are affected by time of day, sound, lighting, temperature, room design, food intake, social environment, and clothing.

Effective Math Study Skills: The Fuel of Excellence

Imagine for a moment you're giving a surprise party for someone special. You've invited people you care about, and you want everything to be perfect. Are the balloons inflated? Does the party table look just right? Are food and drinks set out? Are the flowers fresh? Has just the right music been selected? Don't you think things are more likely to go well if these arrangements are in order? Preparations pay off. And if you want your journey toward your math goal to go smoothly, with less hassle, it's absolutely crucial you set the stage for learning. How? By sharpening your math study skills.

Don't shrug your shoulders. Math isn't the same as other subjects. Strategies to help you study and learn psychology, history, or English won't necessarily help you with math. There are studying techniques that work best for math; master them and master good habits that breed success. Master them, and you will remember math concepts more easily, retain them longer, and use them more effectively.

This chapter explores these study-skills methods. I urge you to approach them seriously. They have already provided many math-anxious students with major breakthroughs on their road to math success. Why not you? At times, these skills may seem disarmingly simple, but believe me, they are more powerful than you think! They are choice ingredients in your math mastery recipe. And remember, using good study techniques won't necessarily mean you have to study more hours; it means you will study more efficiently and obtain gratifying results.

How Good Are Your Study Skills?

EXERCISE 8-1 **My Personal Math Study-Skills Inventory**

This inventory will help you assess the effectiveness of your math study skills. Read the statements carefully and determine how frequently each applies to you. Enter the correct point score for that item (usually = 3, sometimes = 2, and rarely = 1).

	Usually (3 points)	Sometimes (2 points)	Rarely (1 point)
1. I attend all my math classes.	____	____	____
2. I do my math assignment before attending class.	____	____	____
3. In class, I mentally follow all explanations, trying to understand concepts and principles.	____	____	____
4. In class, I write down main points, steps in explanations, definitions, examples, solutions, and proofs.	____	____	____
5. I review my class notes as soon after class as possible.	____	____	____

	Usually (3 points)	Sometimes (2 points)	Rarely (1 point)
6. I review my class notes again 6 to 8 hours later or definitely the same day.	___	___	___
7. I do weekly and monthly reviews of all my class and textbook notes.	___	___	___
8. In reviewing, I use all methods, such as reciting aloud, writing, picturing the material, etc.	___	___	___
9. I study math before other subjects and when I'm most alert.	___	___	___
10. I take short breaks every 20 to 40 minutes when I study math.	___	___	___
11. I work to complete my difficult math assignments in several small blocks of time.	___	___	___
12. I reward myself for having studied and concentrated.	___	___	___
13. I survey my assigned math readings before I tackle them in depth.	___	___	___
14. When I read, I say aloud and write out important points.	___	___	___
15. I underline, outline, or label the key procedures, concepts, and formulas in my text.	___	___	___
16. I take notes on my text and review them often.	___	___	___
17. I complete all assignments and keep up with my math class.	___	___	___
18. I study math 2 hours per day, at least 5 days a week.	___	___	___
19. I work on at least ten new problems and five review problems during each study session.	___	___	___

	Usually (3 points)	Sometimes (2 points)	Rarely (1 point)
20. I work to "overlearn" and thoroughly master my material.	____	____	____
21. I retest myself often to fix ideas in memory.	____	____	____
22. I work to understand all formulas, terms, rules, and principles before I memorize them.	____	____	____
23. I use a variety of checking procedures when solving math problems.	____	____	____
24. I study with two or more different math books.	____	____	____
TOTALS:	____	____	____

To find your total score, add up the total points of all three columns:

MY GRAND TOTAL IS: _____

If your score is above 68 points, you have excellent math study skills.

If your score is between 54 and 68 points, you have fair math study skills, but you need to improve.

If your score is below 54 points, you have poor math study skills, and you need help fast!

Now let's see what you can do to improve your math study skills.

How Do You Approach Learning?

One traditional approach to the learning process is the *empty vessel* model. In this model, the student considers his or her mind an empty container that the teacher is expected to fill with knowledge. If you adhere to this approach, you would expect to learn all there is to learn simply by attending class and opening your mind to the subject matter, which then pours in. I am *not* a believer in this approach. My many years as a college counselor have shown me learning doesn't happen passively.

We are very complicated beings, and we must understand ourselves and how we learn best before we can reach out and tackle a subject. We must understand how we think, create, and retain new information. We must understand the memory process and strategies that enhance recall. We must look at the uniqueness of our subject matter and identify how best to approach each new area of knowledge. This is the *personal growth* model of learning. The user-friendly strategies described here are based on this approach.

Which Math Class to Take?

Let's begin our discussion of math study skills with your plan to take a math class. This is a very important step, and you must execute it carefully. There are three aspects you should consider: (a) correct placement, (b) when you last took math, and (c) auditing or repeating a math course.

Correct placement in math is crucial. If you are put into a math course too advanced for your level of understanding, your anxiety level is sure to increase. The material will be over your head, and you will be doomed to failure. On the other hand, if you are put into a class below your math ability, you may get bored. Contrary to what you might think, when students are placed lower than their math ability level, they tend to do poorly in math and often stop coming to class or drop out. So being placed correctly is the key. If possible, take your school's math placement exam or find a math teacher who can accurately assess your math skills and place you properly.

When did you last take math? I encourage people who are taking math never to skip a semester if they can help it because it's easy to forget a lot fairly quickly. One study found that after 1 year of nonuse, students had lost approximately two-thirds of their elementary algebraic knowledge (Pauk & Owens, 2008). If you've been out of school for many years, you are sure to have forgotten much more of your math. A math placement exam could help you decide which math course to take.

As a rule of thumb, if you haven't taken math in about 3 years, you probably will need to repeat the last math course you successfully completed. And think of successful completion as a grade C or better. In fact, research has indicated students are more likely to succeed as they go into higher and higher levels of math if they have no lower than a B in their algebra courses.

If you decide to repeat a course for credit and you receive a higher grade than you received before, most colleges will use only the higher grade in your grade point average (GPA) and eliminate the lower grade from your GPA. Find out how your school handles repeats.

If, to freshen up your skills, you decide to repeat a course you have already taken, you may consider auditing the course so you don't have to be concerned about the grade. But I urge you when auditing the course to come to class regularly, complete all the homework, and take all the tests. You will get the most out of the course this way, without having to worry about a grade. Auditing can be used in another way as well. If you plan to take a math course that you feel is a difficult one, you might audit it first and take it for credit the following semester. The audit allows you to familiarize yourself with the material, and it gives you the opportunity to practice working out the math problems. However, don't audit unless you plan to participate.

Once you've chosen the correct math class, follow the exact math sequence recommended at your school. Do not skip over a course or skip a semester without taking math. Work sequentially until your math requirement is completed.

Schedule It Right

A truism for many students is: *Learning math frequently and in small chunks is a formula for success.* So how often your class meets each week is an important consideration. I urge you never to take a math class taught only once a week. Always choose a class that meets as many times a week as possible. For most three-credit courses, this usually means meeting three times a week during the regular semester, but I have seen some sections of college algebra, trigonometry, or calculus held four or five times weekly. Remember, the more days a week, the better.

The time of day you take the class should also be carefully chosen. Taking into account the chart you completed in Exercise 7-3, schedule your class for a time when you are most alert and able to learn. If you are not a morning person, don't schedule yourself for early classes. This will not be productive and may create feelings of frustration and defeat.

Do not schedule anything during the hour following your math class. This will allow you time to review immediately after class, reinforcing the concepts you just learned. Equally important, it permits you extra time should your exam run overtime. I've seen many math teachers allow students to continue working on exams even after the class period has ended. Why not give yourself as much time as possible?

Teacher Selection: Satisfaction Guaranteed

Once you have decided which math course you will register for, the next critical step is to choose a good math teacher. Spend time asking around about math teachers. Ask others why they like or dislike a certain teacher. Don't choose a class just because it fits into your schedule well.

A great teacher can make all the difference in how you feel about going to class and how much you learn in class. Select a teacher who explains concepts well; teaches according to your learning style (see Chapter 7); welcomes questions before, during, and after class; has office hours for outside help; has a positive attitude toward students; and gives fair tests.

Unfortunately, in some schools, students are unable to choose their instructor before registering because the word *staff* and not the teacher's name is listed in the class schedule. If you find yourself in this situation, you will need to determine within the "drop/add" period (approximately the first 10 days of school) if your current math teacher is a person from whom you can comfortably learn. If you find yourself dissatisfied, switch quickly to another section with a more satisfactory instructor. Of course, before changing, check out your new teacher. It is up to you to select well.

We have all had teachers whom we remember fondly (or not so fondly) and whose image and manner are inexorably linked in our minds with the subject they taught. Some students actually are so put off by a teacher's appearance or personality that they come to dislike the course and, by association, the subject matter also. By "losing the forest for the trees," they sabotage their progress.

You may also want to know if your teacher teaches more than one section of the math course you plan to take; if you have to miss a class or if you haven't quite gotten a certain concept in class, you could ask to sit in another section of the same course.

Give It Your All

Once you've chosen the appropriate course and the best teacher for you, you are ready to attend a math class. I encourage you to view the learning of math as a positive and rewarding academic challenge. Devote your energy to it. It requires persistence, concentration, discipline, patience, and lots and lots of practice. Don't take math with other hard courses or a busy workload. Give yourself the time! Many teachers and students have learned the "rule" that you should study 2 hours for every 1 hour you are in class. But this may have no basis in reality when it comes to math or any other difficult subject. Successful math students usually study math for at least 2 hours every day throughout the semester. So don't feel bad if you don't assimilate math as fast as you think you should. Learning math takes time. Give yourself the time; nothing succeeds like excess!

Stay Current

It is most important to stay current in math. Don't fall behind or the entire course will become an effort and a struggle for you. Success in math builds on existing knowledge at each stage. Be attuned to the cumulative nature of math. You can only understand new information if you assimilate and digest earlier information. So keep up with the work and don't fall behind; try not to miss important building blocks along the way.

Attend All Classes

Successful students are more likely to attend all classes, whereas failing students miss one-third or more of their classes. Don't cut math class. Missing even one class may actually put you behind by at least two sessions because you may feel lost when you return to class the next session. If you have to miss, be sure to read the assigned text material thoroughly, do your homework, and get a copy of the class notes from a classmate. Go to your teacher to clear up anything you don't understand.

Be Bold: Sit Near the Front

Successful students are also more likely to sit close to the front of the classroom and near the center. Be bold—boldness has genius. You are more likely to pay attention and concentrate on the lecture by sitting close to the front. It also helps you to be more involved in the class, to have more direct contact with the teacher, and to see the board more easily. Those students who sit in the back of the class are not only physically but also psychologically more distant from the lecturer. It is much easier to get distracted by sounds or side discussions going on in the back of the room.

Take Class Notes

Your class notes and your text notes are like your bible in math. They indicate the essence of what you are learning. Studies show successful students take more class notes—about 64% of what is presented—than unsuccessful ones. Write down what the teacher puts on the chalkboard and all verbal explanations that can clarify what you are learning. Write down important ideas, equations, examples, helpful hints, and suggestions. Strive to follow and understand the teacher's reasoning and logic when solving a board problem. Note steps in a solution the instructor explains but doesn't necessarily write on the board. Don't be afraid to ask the instructor to repeat anything you miss or don't understand. If your teacher explains examples directly from your math text, you needn't write these down. Simply follow along in your text and add any clarifying statements. Make your notes legible, neat, and clear so you can read them easily.

Be an active listener. No teacher can speak or write as fast as we can think, so it is very easy to "tune out" during math class. I hear this from lots of students who have trouble concentrating in math class. They begin to daydream, or think about the chores they need to do or about a relationship they're in, and pretty soon, they've lost the train of thought in the lecture. So be an active listener. Relate what is being taught to previous lectures, to the homework assigned, or to your textbook reading. Pose questions to keep up your interest and further your understanding.

Choose a large notebook with pockets you can use exclusively for math. In the front half of your notebook, write class notes, and in the back half, write textbook notes and the solutions for homework and sample practice problems. Use the pockets for handouts, syllabi, or returned test papers. Date each day's class notes and identify the topic on top of the page in large writing. Leave plenty of space in your class notes for additional clarifications, diagrams, sketches, and comments you may want to add later.

It's an excellent idea to use your colored markers when you review your class notes to identify definitions, theorems, proofs, formulas, procedure steps, examples, or equations. If you clearly label your class notes, you will be more efficient in locating information to review and study for tests.

Questions That Count; Answers That Add Up

Always remember, you have the right to ask questions of your teacher before, during, and after class. See your instructor during office hours or visit the math learning center, if your school has one. Notice when you begin to fall behind and seek help immediately. Never avoid asking questions because you are afraid to look stupid. There is *no such thing as a stupid question!*

Design questions for yourself when you read your chapter, when you do your homework assignments, and when the teacher is explaining concepts in class. If you have difficulty following a procedure your instructor or a tutor is working out, ask questions about

"PARDON ME SIR – COULD YOU PLEASE EXPLAIN STEP 3?"

each step you find confusing. Determine which procedural step first caused your difficulty. Then ask a question about this precise step. Rather than say, "I'm lost," ask, "Could you please explain step 3?" or "I don't understand step 3. Would you explain it a different way?" By being specific, you can pinpoint exactly where your difficulty began, and in turn, your instructor or the tutor can be more effective in helping you. They are more likely to avoid repeating information you already know and, instead, focus on your area of difficulty. So speak up, ask specific questions, and talk your way to success!

The Magic of "Now" Improves Memory

I've had many students come for counseling and tell me they understood the material presented in their math class, but when they looked at their notes and homework a day or two later, it was all "Greek." They must then sit for hours reconstructing what it was their instructor taught a day or two earlier in class.

I have found the most important study skill math students can learn is to *review immediately after learning* and then again 8 hours later. This review directly after math class is critical. Reviews need only last 10 or 15 minutes because you already know the material. Cover your notes and repeat them to yourself in your own words or picture them in your mind. Cover up the solutions to problems worked out in class and see if you can work them out now. It is also important to do your homework the same day as your teacher assigns it. This acts as a review of what you just learned. Make a commitment to yourself to review regularly throughout the semester.

Let me explain to you why it is so important to review immediately after learning. The German psychologist Hermann Ebbinghaus was the first person to do research on the rate of forgetting. He did his research with meaningless material known as nonsense syllables. He found that, after 20 minutes, nearly half of what had been learned was forgotten, and after 1 day, nearly two-thirds were lost. Ebbinghaus also found that after 2 days, 69% was lost; in 15 days, 75%; and in 31 days, 78%. This means that, after a month, you remember only 22% of the material you learned.

A classic study by H. F. Spitzer on the retention of meaningful material found results similar to Ebbinghaus's study. Even with meaningful material, most forgetting takes place immediately after learning occurs. However, Spitzer showed that students who reviewed the material immediately after learning and then did periodic reviews were able to retain almost *80%* of the material after 2 months! Why this astonishing difference?

One reason reviewing immediately after learning is so effective is that it takes time for new information to be consolidated or converted into permanent or long-term memory. There is a reverberation circuit of neurons in the brain that helps in this process. If you stay with the material you just digested and review it immediately after learning, this circuit is repeatedly activated until your *memory* or *neural trace* is strengthened, forming long-term memory. Estimates indicate it takes between 4 seconds and 15 minutes for a memory trace to consolidate or jell in your mind. If you don't review immediately after learning new material but instead jump to another subject, watch the news, or run off to another class (as with back-to-back classes), you prevent this memory trace from becoming established in your long-term memory. So stay with it; review soon after learning and often thereafter. You'll find you will start to remember more. And because you know the material, you don't have to spend hours relearning it. This is the most efficient and effective way to learn math. So don't delay; review your material promptly and capitalize on the magic of *now!*

The Birth of Excellence: Say and Do

In the personal growth approach, I encourage you to be fully involved in the math learning process. This means use all your senses: Recite the material aloud; explain it to others; hear it; see it; write it down; work with it; manipulate it in as many ways as possible; work out as many problems as you can. This ensures that, if you can't remember your material one way, you will another. If you use a variety of methods, you will also be reviewing your material and reinforcing your memory of it. This approach to learning is further substantiated by memory retention studies (as reported by Magnesen, 1983) showing that students tend to remember a full 90% of what they "*do* and *say.*"

So jump right in and be involved! Make sure to take both class notes and reading notes. Taking notes is an active process, and the more involved you are in the learning process, the more you will learn. Also, as you read your math book, recite aloud in your own words what you are learning. This is an immediate review, and it will help consolidate the material in your memory.

If you sit in math class and take no notes or silently read your math book without reciting aloud or taking notes, you will remember very little after a few days. We remember only 10% of what we read and 20% of what we hear. But studies show students remember 70% of what they explain aloud to themselves or others. Furthermore, by involving all their senses, by both saying and doing (writing, manipulating the material, working out problems), students are able to remember an incredible 90% of the material learned.

When you go to a math tutor for help and the tutor explains the material to you, who is going to remember more, you or the tutor? The tutor, of course! If the tutor is "saying and doing," the tutor's abilities are sharpened. So if you go to a tutor for help, after the material is explained and shown to you, say, "Now let me explain it to you and work out a different problem to see if I understand."

Homework Hints

I mentioned earlier you should always strive to complete homework assignments the same day they are assigned to ensure your review takes place as soon after classroom learning as possible. Make sure to read your assignments before you tackle your homework. In an effort to save time, many students attempt to do their homework problems without reading about the topic. In the long run, this turns out to be a wasted effort because they miss essential details and must start all over. So be methodical. Work on all the problems you were assigned and even more. The more problems you work out, the more your confidence, competence, and speed in doing math will build. It may seem like a lot at first, but your increasing fluency and self-esteem will reward you handsomely.

When you are doing your homework, it is okay to grapple or "wrestle" with it. Homework is intended to challenge your thinking and test your newly acquired knowledge. And remember, there isn't only one right way to solve a problem. Try at least two or three ways. Determine for yourself which approach works best.

Devote either the latter half of your math notebook or a whole separate notebook to your homework. Carefully label and date each assignment and give yourself plenty of paper to work out each problem. Don't cram your work on the page. By giving each problem adequate space, your teacher or a tutor can help you locate mistakes more readily. It will also be easier for you to review your work later when you are studying for exams.

Visualize the Problems

Read each one of your homework problems carefully at least two or three times. Do you understand it? Can you state it in your own words? Determine what things are given, what the unknowns are, what relationships exist, and what the problem asks for. Write these down. Now ascertain how to achieve the results from what you already have. Look at other problems you've learned and see if the same procedures can be used. Try to locate simpler problems similar to this problem. Draw out the problem. Make tables, illustrations, diagrams, and so on. This will give you direction for your analysis and computation. An illustration often can clarify the meaning of a difficult problem or formula. If you're a visual learner, it will be a double bonus for you. Write the equations needed. Estimate the answer and decide on the operations to be done. Then do the necessary manipulations, checking yourself step by step. Once you are finished, use any of the checking procedures described next. For more information on problem solving, read the suggestions in Chapter 10.

The Ten Commandments of Work Checking

Many students lose a significant number of points on their homework problems and on tests because of errors in simple computations, not because of a lack of understanding. If you want to get the highest score possible, it is important you learn to check your work carefully. This process is similar to proofreading your essays in an English course. Get into the habit of checking all your homework problems as if you were taking a test. It will be good practice for the real thing. Use the following guidelines.

1. *Does your answer make sense?* Is it reasonable? Reread the problem to be sure you have approached it in a logical, systematic way.

2. *Does your answer fit your estimate?* When you work out problems, be sure to estimate the correct answers first. Estimate reasonable upper and lower limits; in other words, find the range within which the answer should fall. Then work the problem out and see if the answer and the estimate are of about the same magnitude. If not, rework the problem in a different way to see if your new answer is close to the estimate.

3. *Recalculate.* Recheck your division by multiplying, your multiplication by dividing, your addition by adding the numbers in a different order, and your subtraction by adding. Many students make simple computational errors in these operations, leading to a significant point loss on exams. Others reverse or forget to carry numbers, forget the middle term when squaring binomials, or don't do the same operation to both sides of an algebraic equation. If you use a calculator for these operations, make it a rule to calculate each operation twice. Always check and make sure you are using the right order of operations when solving algebraic equations. When factoring

using exponents, are you applying exponents to all the factors? Have you multiplied the exponents when raising an exponential form to a power? When dealing with radicals, have you applied the radical to every factor inside the radical?

4. *Do your problem twice.* Some students find it helpful to complete the problem and then to do it over again (using an alternative method, if possible) without looking at their previous solution. If the two answers are different, they know that one solution is definitely incorrect. At this point, they go over each step and carefully check for errors. Other students prefer to check for errors at each step along the way during problem solving.

5. *Check your usage of signs.* When you multiplied two negative numbers, did you get a positive one? Did you change a negative sign to a positive one when you switched it from one side of an equation to the other?

6. *Check your decimal points.* Did you put the decimal point in the right spot? If you estimated the correct answer, it will be easier to check if your decimal point is properly placed.

7. *Recheck your writing.* When you worked out problems on scrap paper, did you transfer them correctly to the answer sheet?

8. *Check your exponents.* Are you handling exponents correctly, multiplying, dividing, adding, or subtracting the necessary values?

9. *Reread visuals.* If you are required to read charts, tables, figures, or graphs, do you double-check everything you are looking for?

10. *Substitute your answer.* Does your answer satisfy the given conditions of the problem? After working out the problem, take the answer you got and substitute it for the unknown quantity in the problem. If it doesn't fit, double-check your calculations or work it out a different way.

Practice, Practice, Practice

Nothing succeeds like excess. Work out lots and lots of sample problems: practice, practice, practice. Make problem solving a part of every study session. This is your most powerful aid to effective math learning. As a rule of thumb, work out at least ten problems per study session and review at least five problems from previous study sessions.

Your proficiency in solving math problems increases with practice. Cover up the solutions for problems solved in your text and work out the problems yourself. Study math every day if possible or at least five times a week. The more you review and work out problems, the more you will learn and the more comfortable you will become.

Overlearn Your Math

Since math constantly builds upon itself, each successive layer of new information must rest firmly on previous layers. As you assimilate more and more math, you must firmly establish the principles and concepts you are learning so they will be solid supports for the knowledge yet to come. Overlearning is an important study skill to solidify your math building blocks.

Constantly test and retest yourself on the material you absorb. Learn to recognize your material no matter the order in which it is presented to you on a math test, no matter how difficult it may seem, and no matter how it may be "disguised." When you study, don't stop after you've just barely received the information. Apply the principles you've learned in different situations; work out mountains of problems. Get to know the concepts inside and out. In addition, it's important to review previous knowledge and topics you've learned. This will ensure prior building blocks remain firmly in place and don't erode with the passage of time.

From Acquisition to Understanding

Researchers Cates and Rhymer have studied the relationship between math anxiety and performance (2003). They found that students with high math anxiety levels aren't less accurate than less anxious students when they solve math problems, but they tend to be less fluent or slower. They also found that math anxiety becomes more apparent when the math required multiple procedures to be carried out and when multiple digits were used.

Cates and Rhymer concluded that math anxiety is more related to higher levels of learning math than to the initial acquisition stage of learning. Their research implies that math anxious students are able to acquire new math concepts and use them accurately, but they do not necessarily master these concepts well enough to use them quickly and effortlessly, nor can they readily apply the newly acquired concepts to other similar problems or adapt the concepts to more difficult ones.

Cates and Rhymer suggest students need to go beyond the initial acquisition learning stage to higher learning stages, such as those referred to by Haring and Eaton (1978) as the fluency stage, the generalization stage, and the adaptation stage. Skill in the last two stages reflects a higher level of understanding.

Let's look at these learning stages for a moment. The first stage is the mere acquisition of information. Here the focus is to learn to be accurate in using the new information as measured by giving the correct answer to math questions using the concept, regardless of how much time it takes. Very often, students consider particular math concepts or skills learned when an accurate answer is given. Many math anxious students get stuck at this stage, and don't go to the higher stages when learning to apply this concept. Do not allow this to happen to you.

The second stage of learning is to achieve math fluency. This is the ability to perform a skill quickly, readily, and with little effort. To be fluent, time matters. Math anxious students often acquire basic math concepts and skills and use them correctly when given enough time. However, when the problems become more difficult and they must work faster, their anxiety gets in their way and performance suffers. I encourage you to learn your math skills so well through practice and through overlearning that you can go to the higher learning stages and, hence, increase your mathematical proficiency and performance.

The third stage of learning is the generalization stage. During this stage, students are capable of taking their learned math skills and using these skills in other similar situations. To be successful at this stage, take the math skills you have learned and practiced and solve as many similar problems as you can.

The fourth and last stage of learning is the adaptation stage. Here students are able to adapt or modify the skills they have learned to solve many different types of problems. For success at this stage, I encourage you to apply what you have learned to a variety of more and more difficult problems. In this way, you will become more proficient in math, and your comfort level will as well.

Figure 8-1, Math Learning Stages and Strategies, summarizes the goals of each of the four learning stages and suggests strategies to be used at each stage.

FIGURE 8-1   MATH LEARNING STAGES AND STRATEGIES (Based on Haring & Eaton, 1978)

Learning Stage	Suggested Strategies
Acquisition The goal of this stage is to be highly accurate when using newly learned math skills.	1. Work for accuracy when learning new math skills. 2. Get immediate feedback on whether you are correct or not. 3. Practice the new skill until you can correctly apply it.
Fluency The goal of this stage is to learn math skills well enough to respond accurately, quickly, and effortlessly, particularly when these skills must be combined with other math skills.	1. Review and practice newly acquired math skills until you can use them rapidly, with accuracy, and without hesitancy. 2. Practice being able to rapidly use a combination of math skills when solving math problems. 3. Give yourself time challenges, such as being able to solve five new problems in 5 minutes. When this is accomplished, decrease the time allotted.
Generalization The goal of this stage is to use the learned math skills in a variety of similar situations and to be able to distinguish when to use these skills as opposed to other similar skills.	1. Practice the skill in a variety of similar problems. 2. Looking at a variety of math problems, distinguish between when these math skills are needed and when they are not. Then work the problems out using the skill.
Adaptation The goal of this stage is to adapt or modify relevant elements of previously learned math skills to solving new, challenging, and different types of math problems.	1. Apply the math skills you have learned to more and more difficult and challenging problems. 2. Continue to challenge yourself to solve more and more novel math problems. Determine which elements of the skills you have learned are needed to solve these problems. Then go ahead and solve the problems and check your answers for accuracy.

Follow Your Alertness Cycles

Using the chart you completed in Exercise 7-3, determine when your energy level is high and plan your study sessions accordingly. I call this "rolling with your alertness cycles." However, many students put their math studying off until all their other work is completed, and by this time, they are often tired. This is like running up a down escalator with your arms full of books. You get exhausted and go nowhere fast! So be sure to study math *before* all your other subjects. By studying math first, when you're most alert, you will be a more efficient learner. The concepts will come to you more readily.

Take Breaks

Many studies have shown the distinct benefits of distributed practice—short study sessions interspersed with rest breaks—over massed practice—one long, continuous session without a rest. In math, several short study sessions of intense work and concentration are much more effective than sitting for hours trying to figure out problems. I recommend you study for 20 to 40 minutes and then take a 5- to 10-minute study break. During your break, stand up, move around, stretch, get something to drink, and then return to your studying. After 2 hours, take a much longer break, perhaps 20 minutes. This regimen of short study periods with small breaks prevents mental, physical, and emotional fatigue and keeps your motivation high. You'll find even during your short 5- to 10-minute breaks, your mind will be working on math, and when you come back to studying, your work will go more easily for you.

Ed and Nick, two brothers studying electronics, came for study-skills help. They said they had studied electronics math for 5 hours together, and at the end of that time, they felt they knew less than they knew at the beginning of the session. I asked them to make only one change in their study habits. I still wanted them to study for 5 hours but to spread their studying into five 1-hour study periods distributed over 2 days and to interrupt their 1-hour study periods with a 5- or 10-minute break every half-hour. A few weeks later, they reported that this made all the difference. Now they were learning and retaining everything they studied.

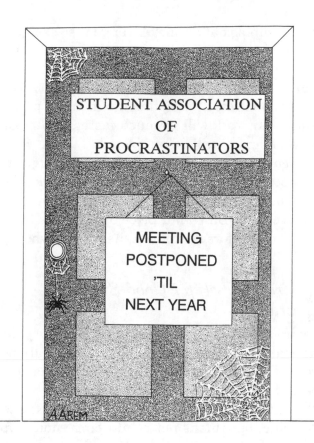

STUDENT ASSOCIATION OF PROCRASTINATORS

MEETING POSTPONED 'TIL NEXT YEAR

Stop Procrastinating: Do It Now!

Are you a motivation killer? Do you like to put off doing today what can be done tomorrow? If you tend to procrastinate and put off your math studying, it's time to take *super action*. Super action is my four-part formula to overcome procrastination. Here's how it works.

Part 1: Keep Taking Little Bites

Many students procrastinate because their math assignments seem too large, overwhelming, or complicated. To deal with this problem, I encourage you to dissect your studying into small "bite-size" pieces. This results in a series of subtasks. For example, you might decide to read your assignment in small sections or separate it into different topics, theorems, or formulas and then practice each separately.

Work on one subtask at a time. Arrange it so each task takes you only a short time to accomplish. If the assignment is particularly hard, spread the tasks out. Take lots of breaks between tasks or complete them in more than one study session. You will accomplish quite a bit if you work on difficult assignments in small pieces or in short blocks of time. Your work will go much faster, and before you know it, the whole assignment will be completed.

Allen, a trigonometry student, found this technique really helped him to stop procrastinating on his math homework. He said he felt like a small mouse who was determined to finish a large hunk of delicious Swiss cheese. Bit by bit, piece by piece, he took lots of little bite-size pieces, and before he knew it, he got the whole job done.

Part 2: Reward Yourself for Good Study and Concentration

Set up a reward system for yourself. It can increase your motivation for studying math and help discourage procrastination. A principle we learn from behavior modification is that if a behavior is rewarded *after* it is performed, it is more likely to continue.

Many students feel their reward is the grade on the test for which they are studying, but this is not reinforcing enough for most students. A test may or may not turn out the way you want it to. I believe more important than the grade on the test is the fact you have studied and concentrated well. So stop procrastination cold! Reward yourself when you accomplish what you've set out to accomplish.

Part 3: Use a Kitchen Timer

Did you ever imagine a simple kitchen timer could stop procrastination in its tracks? It works miracles for people who use one.

If you have trouble sitting down to study, set your kitchen timer for 15 minutes or half an hour depending on how you're feeling. Make sure the timer is in a different room from where you are studying. Now tell yourself: "I only have to study for this period of time. When the buzzer sounds, I'll go and shut it off, and at that time, I'll decide whether I will set the timer for another small block of time." This puts studying totally in your control, and it forces you to move around and take small breaks. Even if you study for at least one

15-minute block, it's a lot better than procrastinating all day. I've also found that once students study in little blocks of time, they become involved in the material and often decide to reset the timer and study more. Try it—you might like it!

Part 4: Change Your Inner Self-Talk

Many procrastinators get stuck in negative, self-defeating self-talk. I urge you to tune into this inner dialogue so you can challenge it with super action self-talk. Here are examples of shifting this negative dialogue to action-oriented self-talk. After reading the eight examples of super action shifts in Exercise 8-2, add at least two of your own.

EXERCISE 8-2 Shift to Super Action Self-Talk

Procrastination Self-Talk	Super Action Self-Talk
1. I don't know where to start.	I'll divide my assignment into small chunks and work on only one chunk at a time.
2. I don't know how to do it.	I'll look up the information in other math books and see how they explain it. I'm bound to get it!
3. I'll wait until I can ask the teacher in class.	I'll work on some of it now and then I'll have better questions to ask.
4. I'm just not in the mood to do it.	Do it now! Just do it!
5. Oh, I still have time; I can wait until later.	Time has a way of running out quickly; get it over with *now!*
6. There's just too much to do.	Take one step at a time; Rome wasn't built in a day.
7. I feel bad, but I keep wanting to put this off.	If I do even a little bit of work on this, I know I'll feel better.
8. I prefer doing my favorite subject first. I can always do math later.	Once I get into my math, it'll begin to be more enjoyable and fun.
9. (add your own)	(add your own)
10. (add your own)	(add your own)

So make up your mind to get things done *now!* It's a mental blueprint for success, not just in math class but in every job or career. Employers love what super action accomplishes. It's reinforcing, and you'll get hooked on it, too.

Tackle Your Math Book

Paula, a young woman taking a developmental math course, came to me because she was struggling in the course. As we talked, Paula confided she had never read her math book. She said she depended on the teacher's lectures to give her all the information she needed for the exam. What a major mistake! By not reading her math book, she missed crucial information highlighting what was being taught in class. The more ways you can be exposed to math—the more you read, the more problems you work out, the more you do—the more you'll learn.

I also encourage you to read your textbook's preface early in the semester. The preface provides an overview of what to expect in your course as well as insights into the textbook author's approach.

Read the Chapter Before and After Class

Make it a practice to read over the assigned chapter *before* you attend math class. Many students avoid reading their math book or wait until after they've heard their instructor's explanation before they read their assignment. If you are one of these students, take heed: Read your chapter prior to class even if you have little understanding of it. Look over the topics, diagrams, charts, terminology, formulas, and examples. This familiarizes you with the topics your teacher will present in class, and it prepares you to understand and absorb the material presented in class more readily. Pose questions on any confusing or difficult ideas in the chapter that can be answered in class, and then listen carefully to your teacher's explanation. You'll find you become a better listener, and you'll learn much more from the class than previously. After attending class, first read your math assignment in depth, and then do your homework problems. This will further your understanding of the math topics discussed in class and will also be an excellent review.

How to Read Your Math Text

Math reading assignments need to be tackled at least three times. Before you panic, let me explain; it's amazingly simple and straightforward. First, survey your assigned material. When you survey, read the lead paragraphs, the first sentence of each paragraph throughout the section, and the closing or summary paragraphs. Read all highlighted areas, tips, subtitles, illustrations, charts, and graphs. This will give you a basic idea of what the section is about. As you do this, pose questions you believe your text material attempts to answer or you believe might be answered in your math class. Be sure to complete this survey before you go to class.

Your second reading is your in-depth one, when you'll read for mastery, reread sections for understanding, and mark and underline in your book. Have you noticed reading your math book is quite different from reading texts in psychology, the humanities, English, or anthropology? In these other areas, as much as 80% of the actual words are unimportant, except for the fact they are needed to link relevant ideas together. When you read math, *every word is important.* Math books are usually written succinctly and to the point. Each word is carefully chosen to explain a concept. You need to read a math text slowly, carefully, and with good concentration. Don't rush. Work on grasping each concept before going on. Learning math takes time. Reread the concepts, several times if necessary, until you gain mastery. If, after several attempts, you still can't understand a particular sentence, review the topic up to this point and then reread the sentence. You also can formulate questions on these difficult points that you can later ask your instructor or a math tutor or look up in other math texts.

In other subjects, you may be able to jump around in your text, skip chapters, or read only parts of chapters and still understand the material. In math, each idea builds upon another; each topic presented assumes you've mastered previous topics. You can't jump into the middle of a chapter or skip a few chapters and expect things to make sense. So move progressively in your math text from front to back, mastering topic by topic.

A picture says a thousand words—as do diagrams, charts, illustrations, and figures. *Don't skip over them.* Work to comprehend what they are illustrating.

Write and recite as you read. Some students hesitate to write in their math books which, they fear, would then lose resale value. But many bookstores don't care if the books contain handwritten notes or underlining, and you could always do this in pencil. Let me assure you, the best math students always mark their texts! Writing as you read gets you more involved in the reading process. Remember: We retain only 10% of what we read but 90% of what we say and do. Make notes to yourself directly in your text. Whenever you encounter new symbols or terms, look up their meaning and write them down in the margin. It's important for you to know your math terminology. Label or mark in the margin important ideas, formulas, and procedures. Underline main points. Write out questions you have about the material. Cover up important topics and then explain them aloud in your own words to yourself or to someone else. Keep testing and retesting your understanding and retention of each topic as you read. View your text as one of the main sources for the types of questions you are likely to have on your math exam.

Examples are an integral part of a math assignment. Go over the steps methodically until you understand the terms used and be sure to follow the reasoning used for each procedure. Afterward, cover up the solutions and visualize, say aloud, or carefully work each problem out step by step, by yourself, to see if you conceptualize the procedures. When you finish, compare your steps to the sample steps in the text. Design other little tasks for yourself as you read. Work out all proofs and derivations. Work out as many problems as you can to ensure you assimilate the topics being presented. You will find it easier to learn facts when you comprehend them rather than follow them by rote.

The third reading of your chapter is for review and taking notes. Write down all important ideas, facts, equations, theorems, examples, and summaries in your notebook or on index cards. Work out your homework problems and do lots of practice problems.

Don't feel overwhelmed! When you learn to drive, there's a lot to attend to and some bumps along the way. But practice brings familiarity, then a sure touch, next comfort with increased speed and precise handling, and then smooth sailing. The same will happen with math.

Keep a Math Terms and New Ideas Log

A log is a great place to document and reinforce new knowledge. You can keep the log in a separate section of your math or homework notebook. Label one section "Math Terms" and another "New Ideas."

Each time you learn a new math term, enter the term and definition in your log. In many ways, learning math is like learning a new language. In fact, it's been claimed that beginning algebra has more new terms than first-year French. As in learning any foreign language, once you know the meaning of the words, learning comes more easily. So starting today, build your math vocabulary.

In your log's "New Ideas" section, define, in your own words, each new idea you learn. By putting the idea in your own words, it truly becomes yours. This is part of "doing" and "saying." This is also being an active participant in the learning process. Your log documents what you have learned. By reviewing it often, you will also reinforce your new knowledge. As your log grows larger, you can be proud of your progress.

Flashcards for Quick Recall

Many students find $3'' \times 5''$ cards very practical for reviewing or committing to memory important bits of information. The cards are handy, cheap, and can be carried around conveniently. While at a stoplight, waiting for a bus, or between classes, students test themselves on their homemade flashcards. This practice helps them to review often and to spread learning out. Both are good learning principles. Students using $3'' \times 5''$ cards report they are able to quickly memorize rules and principles without having to sit at a desk for hours studying and drilling themselves.

To memorize a rule or principle, make sure you completely understand every part of it. Then write a sample equation to show how it's used, or illustrate what it represents with a picture. Once you've done this, read the definition aloud about five times, in your own words if possible. Then write the definition on a $3'' \times 5''$ card so you can review it often.

Use $3'' \times 5''$ flashcards to study math's own unique vocabulary. You also can use flashcards for new facts, formulas, properties, proofs, or procedures you wish to recall quickly. Make sure you understand and know how to use all of these before you commit them to memory. If you understand the concepts behind the formulas and various procedures, you will be able to re-create them easily should your recall falter. *A word to the wise:* Sometimes you may have to commit to memory an extremely difficult concept you do not understand. Don't be discouraged. This even happens to students in advanced math courses. Your understanding may come later by asking questions of your instructor or tutor, looking in other texts, and in some cases, only by taking more advanced math courses.

Use Additional Resource Materials

Whenever possible, use the additional resource materials provided by the publisher of your math book. These resources are integrated with the topics presented in your textbook so you can easily locate the concept you are studying and needn't search around.

Be sure to buy your textbook's student study guide. The study guide will provide you with lots of problems on the concepts you are learning. If the publisher also provides CDs, videos, or a Web site, check them out.

It is also very helpful to use two or more math textbooks when studying. They provide you with many more worked-out problems to study as well as numerous sample problems to work out. Make sure you have the answers listed in the back of the book or in a solutions manual. In addition, by having more math books to study from, you have different explanations for the concepts you're learning. Remember, your instructor has probably studied or taught out of several books and will be using this knowledge to teach the course and test you. You also might choose books that appeal to your perceptual learning channels. For example, if you're a visual learner, you might choose texts with colored diagrams, illustrations, or charts. If you're a verbal learner, you might prefer a book with more explanations, and if you're a kinesthetic/tactile learner, you might search for texts with real-life examples with which you can experiment.

Elena, a college algebra student, said when she studied, she spread out four algebra books around her, and it felt as if she had four different teachers explaining algebra to her.

By having several different books, you can get an understanding of the topics you are studying from many different viewpoints. And if your text doesn't quite make sense to you, another might explain the topic more understandably. In addition, by working with different textbooks, you review a topic each time you read about it. Some students find books like the *Schaum Outline Series* useful for obtaining a large collection of sample problems and answers.

Get Help Fast

Don't delay. If you need assistance, seek it immediately. Procrastination is a student's greatest enemy. As soon as you have difficulty with a new concept or problem, get help. Set up a help session with a tutor or a teacher. Never let yourself perpetuate inaccurate procedures or incorrect problem-solving processes.

Here are some suggestions on how to benefit most from a help or tutoring session:

1. Prepare for the help session by writing, ahead of time, specific questions to ask during the session.
2. After your tutor or teacher explains a new or difficult concept, repeat it in your own words to see if you understand it and to reinforce or strengthen long-term memory.
3. After the tutor or teacher shows you how to solve a problem, you, in turn, should work out several similar problems. Each time you believe you understand a new or difficult math concept, demonstrate to the tutor or teacher your newly obtained

knowledge. You, and not the tutor, must always be the one to "say and do." This will ensure understanding and easy recall.

4. After the help session, reinforce what you have learned by doing lots and lots of practice problems.

In addition to getting help from your teacher or a tutor, you might check the Internet. There are several tutorial services and math help Web sites. One of my favorite Web sites is Ask Dr. Math. This is a wonderful Drexel University School of Education Web site where you can ask math questions. Over 300 volunteer "Math Doctors" from all over the world answer questions posed to them by students. Answers are sent back by e-mail. You can find Ask Dr. Math at http://mathforum.org/dr.math.

Study Buddies

Studying with a friend or in a group can be both fun and beneficial. Partners can help you learn new material, reinforce difficult concepts, test your knowledge, challenge your thinking, identify areas of weakness, give you emotional support, and reduce the feeling of isolation when studying.

The best way to benefit from studying with others is to:

1. Choose a partner or friends who are serious about learning math.
2. Set a regular time each week to study, preferably at least twice a week for about 1 hour.
3. Study in an area where you will not be interrupted or distracted by others.
4. Have a chalkboard or large pad available to work out problems.
5. Spend most of the time working out problems and explaining to your partner(s) how you solved the problem.
6. Always be a critical thinker, carefully evaluating, analyzing, and critiquing your work together.
7. Be honest with your partner when you don't understand a concept.

8. Test each other often, especially on all new and/or difficult concepts.

9. Don't allow yourselves to get stuck on any one problem. Mark it so one of you can ask your teacher or a tutor for help with it.

10. Stay positive and supportive of each other.

11. Spend only the first 5 or 10 minutes for socialization and then get down to work. Focus and concentrate your efforts on the concepts to be learned and reviewed.

12. Work out lots of practice problems.

13. When you are dealing with a difficult concept, make sure each of you can explain it to the other.

14. Have fun!

Important words of caution: Avoid studying with others who put you down or make you feel dumb or stupid. This will have a detrimental effect on your learning. Also do not study with others who aren't serious about learning, want only to socialize, complain continuously about math, have such poor math skills they can't challenge your thinking, or stand you up and don't show up for scheduled meetings. When in a study group, avoid "groupthink," or the tendency of group members to maintain group agreement, even at the cost of critical thinking. If you think something is wrong, speak up, check it out, ask your teacher or tutor for help. Don't perpetuate false thinking.

Clear the Minefield

Without realizing it, you may be setting up a minefield for yourself. This minefield is booby trapped with faulty beliefs about your own study process. For example, you might tell yourself: "There's too much math to study and not enough time"; "I just can't get the concepts and formulas to sink in"; "This material is so confusing; I haven't a clue as to where to begin"; "There's just too much to remember"; or "I understood this math procedure in class, but now I have no idea how to use it."

These are just a few examples of booby traps you can unwittingly set for yourself. These traps are a prescription for disaster. They can undermine your studying, hamper your progression, and increase your anxiety level. If you have the belief that cramming before a math test keeps the material fresh in your mind, you may end up not giving yourself enough time to understand your material or to practice your newly learned skills. Alternatively, if you believe you have too much to study and not enough time to get it done, or there's simply too much to remember, your anxiety level might skyrocket, and you may be paralyzed into inactivity. Each mine or faulty belief in your minefield can stop you in your tracks, take you out of the action,

FIGURE 8-2   DISARM YOUR MINEFIELD

The Mines (faulty beliefs)	Some Suggestions for Disarming Mines
There's too much math to study and not enough time.	1. Ask your teacher to identify the most important concepts on which to focus your studying. 2. Begin studying early and give yourself lots of time. 3. Skim over your class notes and your textbook chapter. Get an overview of what the material covers. Study the charts, figures, examples, and formulas. 4. Identify the major topics on which you need to focus and then divide your workload into small, manageable areas.
I've studied and understood my math, but I just can't get it to sink in.	1. Associate the new concepts to math concepts you already know. Determine how similar or different they are from the previous ones. 2. Practice using the new concepts and procedures in all types of problems. 3. Keep testing and retesting yourself. Get immediate feedback on whether you were correct. Then analyze where your errors are so you can correct them with further practice.
This material is so confusing; I haven't a clue as to where to begin.	1. Identify all the key concepts and procedures you need to study. 2. Prioritize the most important areas on which you need to concentrate. Begin working on these areas first.
This material just puts me to sleep.	1. Study with a partner or a study group. Give each other challenging practice tests to make sure you know the material. 2. Find other books that explain the material in a more inviting way. 3. Start studying math early in the evening, before other subjects.
I understood this math procedure (or concept) in class, but now I have no idea how to use it.	1. Review and practice the math procedure (or concept) immediately after class and again later that day and later that week. 2. Practice using the procedure (or concept) with many different math problems.

FIGURE 8-2   (continued)

The Mines (faulty beliefs)	Some Suggestions for Disarming Mines
I can't remember all this stuff; it's just too much.	1. Write summary sheets or flashcards of the important procedures and formulas or put them on handheld computers. Carry them around with you and test yourself often. 2. Use lists, charts, or diagrams to help you best organize the material you want to remember in a way to readily retrieve it when you need it. 3. Create your own creative mnemonics (memory aids) like a rhyme or phrase to help you associate the concepts with something you already know. The goofier or funnier the mnemonics, the better you will remember the material.
I think I might know this math concept well enough to be tested on it.	1. Be sure to test yourself with many different examples to see if you truly understand the math concept. 2. Test and retest yourself until you're absolutely sure you know the concept.
Math goes in one ear and out the other.	1. Push distracting thoughts out of your mind so you can focus on learning your math. 2. After reading about a new math concept, see if you can explain it in your own words. Then practice applying it in solving problems. 3. Get help from a tutor when learning new math concepts. Ask lots and lots of questions to keep you focused and on track. 4. Set up a reward schedule for yourself. For example, give yourself a small reward for every half-hour of good math studying you accomplish.
I'll just cram before the test, and then the material will be fresh.	1. Start studying and reviewing for an exam 1 or 2 weeks ahead. 2. Spaced learning increases retention and recall. Study in small chunks of time in the weeks before the exam. 3. Be rested and relaxed before an exam. Get a good night's sleep. Avoid being physically and mentally exhausted.

and hurt your progress toward your math goal. Figure 8-2 identifies some of the mines that may litter your minefield and offers suggestions for disarming the mines.

Helpful Strategies for Dealing with Dyscalculia

If you're experiencing dyscalculia (see discussion in Chapter 2) many of the strategies discussed in this chapter and in Chapter 7, Enhance Your Learning Style, will be beneficial. Also, check out The Learning Disabilities Association of America's Web site, www.ldanatl.org, which offers some excellent ideas. It suggests you use colored pencils to differentiate different types of problems; draw pictures when working out word problems; use diagrams to increase your understanding of math concepts; devise mnemonic devices to learn steps in a procedure; use your fingers, scratch paper, or graph paper; or listen to music and, in particular, rhythms to help you learn math facts.

If you have difficulty dealing with math sequences, as some people with dyscalculia do, it is helpful to visualize the steps in a procedure. Picture, in your mind, the numbers, the formulas, and the operations you must undertake. Picturing the sequences allows you to keep things in the correct order. When you solve a problem, you usually have to do the steps in a specific sequence to achieve the correct answer.

It is also useful to draw flow charts or diagrams to help visualize sequences of steps. Try using colored pens or markers to highlight various parts of a problem or a procedure. Some students find using graph paper or spreadsheets along with colored writing utensils helpful to map out their work or to organize their ideas on paper. When working with multistep problems, divide them into finer and more manageable steps.

When learning a new math skill, flashcards often help. Be sure to study several concrete examples of how the skill is applied. Then gradually move from the concrete to more complex and abstract ways to apply the information. This will help you understand how to use the concept and not merely memorize it. Practice how to estimate the answer of each new problem before you solve it.

Also, when you study your math, think of it as a second language. Write down and define new terms and symbols. Study math terminology. Knowing the language of math and the meaning of each term will take the mystery out of the terms being used.

Some people with dyscalculia get distracted easily. If this happens to you, make sure to study in a place free from visual and auditory distractions. You want to keep your focus only on the math at hand. You can also use a line reader to help you stay on track and distinguish one line of a math problem from another.

Figure 8-3 offers some suggested strategies to deal with several dyscalculia difficulties experienced by students.

FIGURE 8-3   SUGGESTIONS FOR DYSCALCULIA DIFFICULTIES

Dyscalculia Difficulty Experienced	Some Suggested Strategies
1. Remembering steps in a procedure	Use flashcards; visualize the steps; use mnemonic devices; listen to a rhythm that might help recall it
2. Recognizing patterns in a common procedure or operation	Picture the numbers, formulas, and operations in your mind; diagram or draw the procedure
3. Organizing problems	Use graph paper or spreadsheets and colored pencils or markers
4. Using steps in a math procedure	Draw a flow chart or a diagram
5. Using multistep procedures	Divide the procedure into finer steps and work out each step before going on; use colored pencils to differentiate between parts of the procedure
6. Understanding concepts	Apply the concepts in many examples from very concrete ones to more and more complex, abstract ones
7. Putting language to math processes	Write down new math terms and symbols and define them and study them
8. Understanding and doing word problems	Draw pictures, tables, or diagrams to picture what the problem is asking for; estimate your answer beforehand and plug the answer into the problem to see if it makes sense
9. Staying focused	Study in a place free from all distractions; use a line reader to distinguish one line of a problem from another

Now It's Time for You to Decide

EXERCISE 8-3 **Collaborative Learning: My Commitment for Improvement**

List the five most important study skills you've learned in this chapter that you're willing to incorporate into your study routine within the next month. Share your list with a small group of your classmates.

1.

2.

3.

4.

5.

Summary

Use appropriate math study skills and you'll achieve major break-throughs on your road to success. This section of our road map highlighted strategies for obtaining greater mastery and retention of mathematical concepts. I encouraged you to jump right in and get actively involved in the study process; to carefully select your math class and your teacher; to go to class regularly; to take complete class notes; to stay current and not fall behind; to take the initiative to ask questions; to review immediately after learning and again 8 hours later; to tackle your math reading assignments in three different ways; to work out lots and lots of practice problems every day; to keep a math terms and new ideas log; to use flashcards; to study with a buddy; and to clear your minefield of faulty study beliefs. I provided helpful strategies for dealing with dyscalculia. I encouraged you to go beyond mere acquisition of math concepts to math fluency, generalization, and adaptation.

Conquer Test Anxiety

Too many students experience test anxiety. Fear and overwhelming doom come over them like an impending, life-threatening storm. Darkness fills the horizon, and they shudder in their footsteps, paralyzed, unable to take any action. Thoughts race from one thing to another and, suddenly, blankness, emptiness, all they knew—gone! An empty slate with no numbers, no formulas, nothing. Foreboding images of agony or sudden death loom heavy in the air. Dreams and ambitions melt into oblivion. Feelings of worthlessness, uselessness, and hopelessness abound. Has this ever happened to you? If so, you are not alone. Each day, thousands face the devastating effects of test panic.

The symptoms of test anxiety may be as pervasive as those just described or as minimal as forgetting an easy formula you've used hundreds of times. Either way, your ability to perform well on math exams becomes terribly hampered. In some cases, you may become totally immobilized.

EXERCISE 9-1 Symptoms of Test Anxiety

The following checklist will alert you to the physical and mental signs of test anxiety. Check off any of the symptoms, mild or severe, you have experienced before or during tests.

Physical

_____ increased sweating

_____ increased need to urinate

Mental

_____ confusion, disorganization

_____ foggy thinking

Physical	Mental
_____ headaches	_____ blank mind, freezing up
_____ shakiness	_____ overwhelming fear or panic
_____ upset stomach	_____ poor attention span
_____ pounding heart	_____ poor concentration
_____ loss of appetite	_____ increased errors
_____ tightness of muscles	_____ fleeting thought processes
_____ stiff neck	_____ narrowed perceptions
_____ backache	_____ immobilized creativity
_____ fatigue	_____ nervous worrying
_____ "free-floating" anxiety	_____ pervasive negativism
_____ insomnia	_____ weakened logical thinking
_____ total mental fatigue	_____ feelings of impending doom
_____ feelings of inadequacy	_____ distracting thoughts

Kirk, a college freshman, came to me and pleaded: "Why me? I studied and studied. I knew my information, but I just blew it. I feel so dumb. I confused formulas, multiplied 2 × 4 and got 9, and on top of it all, copied answers incorrectly to the answer sheet." Why, indeed?

The roots of test anxiety are threefold, and any or all may be the culprit. They include poor test preparation and test-taking strategies, psychological pressures, and poor health habits. In this section of our road map, we explore how to deal with each of these troublemakers.

Test Preparation and Test-Taking Strategies

Many students feel anxiety about taking math exams because they have poor study skills and inadequate test-taking strategies. The previous chapter discussed effective math study skills. Good skills are essential for comprehensive and long-term recall in math. In addition, specific test preparation and test-taking strategies make a world of difference. They help relieve pretest jitters and insecurities.

Here are some important strategies I recommend to students.

1. When a test is announced, check in advance its format and content. Know what your instructor expects. Make a detailed list of all the topics the test will cover. Be sure to find out how many questions it will include and the time allowed. Will you need a calculator, charts, tables, or scrap paper? For example, Kamala learned her statistics final was going to include 60 questions, and she would have 3 hours to complete the exam. Each question would have equal point value, so she'd have 3 minutes to answer each question. She learned even if some of her answers were wrong, she could get partial credit if she showed her work. She found out what six major topics the exam would cover. She'd be allowed to bring lots of scrap paper and her calculator. Statistical tables would be provided for all students.

EXERCISE 9-2 **Your Next Math Exam**

Can you answer each of the following questions about your next math test?

1. How many questions will be on the test?
2. How much time will be allowed?
3. Will each question have the same point value?
4. What topics will be covered in the exam?
5. Can you bring a calculator, scrap paper, or other aids?

2. Review each concept the exam will cover. Locate problems illustrating these concepts. Learn how to recognize these problems when they're placed in random order, are worded differently, or the concepts appear disguised. Solve lots of difficult problems on each topic to be tested. Work on gaining mastery of each math concept you're studying. Reexamine all procedural steps used to solve specific types of problems. For example, if you will be asked to solve quadratic equations, plan to review the steps needed to solve these problems: simplify, write in a standardized format, factor (and if not factorable, use the quadratic formula), set factors equal to zero, and then solve the resulting equations.

I encourage you to know your information inside and out so you can practically do it in your sleep. Then, no matter how the information is presented on the exam, it won't throw you, and you'll recognize it no matter how it's camouflaged.

3. After completing your review, design your own sample practice exams covering each topic you've mastered. Choose questions from your text, other math texts, tests from past semesters, study guides, math software, or college outlines series books. Make sure that you have available to you the correct answers for these questions. Plan practice exams that include the same number and type of questions as your test. Design your practice exams carefully so they accurately reflect what has been done in class, what your teacher expects, and areas in which you may need more work.

4. Set a time limit for your practice exam, giving yourself the same amount of time allowed for the test. Then take several timed practice exams. Time yourself with a kitchen timer or an alarm clock. Many students can figure out all sorts of complicated problems if given enough time, but then they panic under the test's time pressures. So be sure to set up your own timed exams. And practice, practice, practice. Work on building up speed and accuracy. If possible, take these practice exams in a room similar to the testing room.

This type of test preparation is great at relieving the stress of the real test and prevents panic. When you finally do take your exam, you can say, "What's the big deal? I know I understand my work. It's just another day and another test. I've already tested myself a zillion times. I'm ready—sock it to me!"

EXERCISE 9-3 Design Practice Math Exams

1. List the sources you will use to get sample problems.

2. How will you get the correct answers to each problem you select?
3. How many questions are you going to include in your practice exams?
4. How much time will you allow yourself to complete each question?
5. Where will you take your practice exams?

5. Use your practice exams to begin to recognize problems out of the context of your textbook or class notes. Note the types of ques-

tions causing you difficulty. Give yourself more practice in these areas. Ask your teacher for further help or go to tutorial or help sessions.

6. Use your practice exams to analyze what typical errors you are making. Identify whether there's any pattern to the errors. For example, do you consistently use one of your formulas incorrectly, or do you forget to carry numbers during calculations? Maybe you put the decimal point in the wrong place, or use signs incorrectly, or reverse numbers when recopying them. Perhaps you multiply instead of add exponents when multiplying exponential forms with like bases, or maybe you confuse the order of operations when solving algebraic problems. Make a list of these errors. Your list will be very important later when proofreading your math test.

EXERCISE 9-4   Your Typical Math Errors

Do an error analysis. Look over your past math exams and your practice exams and list the typical careless errors you have made.

---

7. Get plenty of sleep in the 48 hours prior to your exam and eat healthy meals (see the next section on preperformance health tips). Make sure to practice deep breathing techniques in the half-hour before the exam. Arrive at the exam just on time and discuss the exam with no one. This avoids "catching" others' preexam jitters.

8. As soon as you receive your math exam, write on the top corner of the exam paper *all* the formulas, rules, and key information you'll need, including the simplest ones. Do this first, before beginning your math exam. This way, you'll have a handy reference guide—your "cheat sheet"—in front of you, and you won't have to keep it stored in your head. Most students find this to be a very helpful technique that allows the exam to go smoothly.

EXERCISE 9-5   **Your Math "Cheat Sheet" (*Optional*)**

List all the formulas, rules, principles, and key ideas you will need to remember for your next math exam.

---

9. Note the time limit and the point values for each question, particularly those that carry more value. Adjust your time accordingly, allowing more time for questions with higher point values.

10. Circle or underline significant words in each problem. Read and reread each problem slowly and carefully so you don't misinterpret anything.

11. Start the math test with the easiest problems first. These may not necessarily be the first ones on the test. Do whichever ones come easily for you. You need not go in order. Then go to the next easiest, then the next easiest, working your way on up to the more difficult ones as your confidence level increases. This strategy reduces anxiety and gives you feelings of success.

12. Don't spend too much time on any one problem. Skip the harder ones, but mark them so you can find them readily. Return to these difficult questions later as your confidence builds. Also, be careful not to get stuck for a long time trying to remember a step on any one problem, even if it seems easy. Don't allow yourself to lose too much valuable time on any individual problem that would prevent you from completing the exam. Go on and the key step you are trying to remember will probably come to you later.

13. Throughout the exam, focus on remaining calm, relaxed, and positive. Check your breathing often; keep it regular and slow. Check your neck and shoulder muscles and loosen tight areas. Say positive statements to yourself and push away any disturbing or distracting thoughts.

14. If you can use calculators, take advantage of them. Learn how to use them well ahead of time. There are appropriate and inappropriate times to use them. They can be very useful for checking your answers.

15. When working on a difficult problem, write down whatever you know. This often helps in solving the problem. Some teachers give partial credit in recognition of what you know. Don't skip a question if you know even a little. The answer might come to you as you work on it. But save these more difficult questions until the end of your test.

16. Take the *entire* test period to finish the exam. Don't allow yourself to get nervous when some students leave early. Teachers have found math students who leave the earliest are often the ones who do poorly.

17. After finishing the test, use the remaining time to verify your solutions. First, you should check for reasonableness. Ask yourself: Does the number fit? Does the answer make sense? For example, can Marvin's little sister really weigh 658 pounds? Or did Martha realistically spend $12,490.00 on a week's worth of groceries? Second, check the specifics of the problem. Have you satisfied all the conditions? If you are unsure of your answer, do it again. Use the ten commandments of work checking described in Chapter 8.

18. Proofread your test. Did you answer all the questions? Are there any parts you missed? Look for omissions. Check to see if you've made any typical errors or pattern of errors you identified previously (see Exercise 9-4).

19. Leave and reward yourself for a job well done!

Deal with Psychological Pressures

Psychological factors contribute dramatically to test anxiety. Students often talk negatively to themselves, put themselves down, and experience low self-esteem. No wonder they feel ill-equipped to confront difficult, high-stress situations. Whether realistic or not, some students create high expectations for themselves and fear failure or poor grades will jeopardize their entire future. Family members or loved ones, often unintentionally, may add to this burden with excessive demands to do well. Encumbered by enormous self-imposed and externally imposed pressures to succeed, students are likely to panic at the thought of what might happen if they don't.

Michael was one such student. He gave up his well-paying electronics job to return to school and finish the engineering degree he

FIGURE 9-1 ANXIETY/PERFORMANCE GRAPH

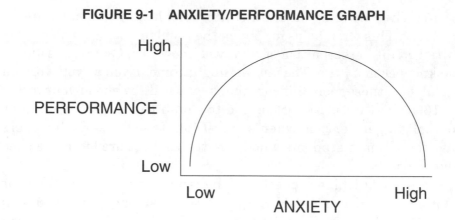

had started 12 years earlier. His wife became the sole breadwinner, supporting him and their three children. Everyone, including Michael's parents, was excited for him and expected him to be a great success. But after three semesters in school, Michael became very discouraged. He was repeating one of his math courses for the third time. It was at this time he came to me for counseling. He had just panicked on an important exam and failed it. Michael appeared to be an intelligent, well-prepared, mature student, but his anxiety level had skyrocketed. He felt like a failure. He was sure he was letting down his family.

Research shows performance is a sensitive barometer to the level of anxiety students experience, though the relationship between performance and anxiety is not linear. You may be surprised to learn the effects of anxiety are not uniformly negative. In fact, research repeatedly finds some anxiety is necessary for optimal performance. When plotted in the form of a graph, the anxiety/performance curve takes the shape of an upside-down U (see Figure 9-1). If students feel low anxiety, they are inadequately motivated, accomplish little, and perform poorly. As anxiety increases to a moderate level, perceptions sharpen. Students feel alert, energetic, clear-minded, motivated, and at their creative best. Not surprisingly, performance reaches an optimal level. But if anxiety continues to rise to higher and higher levels, the effect is catastrophic. Students become increasingly indecisive, make careless errors, and show poor judgment. Overwhelmed by anxiety, students become disorganized, and they can neither function effectively nor think clearly. Performance plummets until, at the highest levels of anxiety, they freeze up, their minds go blank, and total panic sets in.

Once begun, this process is self-perpetuating and extremely difficult, if not impossible, to reverse. Students caught in this bind may have no choice but to abandon the situation. One student I worked with said he totally blanked out on his exam and sat for 30 minutes in a state of panic. He finally left and decided to go home. As he opened the door of his car, all the information he needed for the test came flowing into his mind. He could see all the formulas, all the answers to the problems he had just been staring at. It was as if the doors of memory opened, the floodgates burst, and everything came rushing in. But it was too late!

The key to sustained successful performance on exams is keeping your anxiety within a moderate, manageable level, thus preventing it from ever reaching destructive proportions. You may do this by using a three-pronged approach: (a) reverse negative self-talk, (b) use relaxing and calming techniques, and (c) practice visualization.

Reverse Negative Self-Talk

The first target in test anxiety reduction is the most devastating of all psychological culprits: negative self-talk. I have found students who suffer test panic repeat a continuous stream of negative, self-defeating statements to themselves before, and especially during, an exam. Here are some examples of damaging statements students say:

I can't do it.
I'll never finish all this in time.
It's too difficult for me.
I feel so dumb when it comes to math.
Everyone else knows how to do this but me.
I just don't have the ability.
Who am I fooling? I don't know what I'm doing.
I've had trouble with math all my life.
I hate tests.
I hate this course; I wouldn't be here if it weren't required.
Why do I need to learn this anyway?
I'm too nervous to concentrate.
Maybe I should just drop this course.

The teacher's sure to ask me things I don't know, and I'll blow it
   again.
Who cares about this test [course] anyway?
I always do well on the homework, but the test is never like the
   homework.
The problem looks too easy; I probably got it wrong.

Statements like these tend to become self-fulfilling prophecies.
Students focus their attention and energy on this continuous stream
of inner dialogue. They think less clearly, concentration diminishes,
attention wanders, and they have less energy for dealing with difficult
questions on the exam. More and more, they become distracted by
disturbing or intrusive thoughts. Daydreams about being in a more
pleasant environment—any other place than where they are—are
common. Slight noises or movements in the exam room soon become
monumental distractions. Anxiety levels shoot way out of control.

How can you deal with these negative self-statements? The most
crucial step is to become aware of the negative mind talk and its
ability to sabotage your efforts for success on an exam. Most students are not even aware of their negative inner dialogue, let alone
the devastating effects it has on their performance.

EXERCISE 9-6 **Become Aware of Continuous
Inner Dialogue**

Tune in to your thoughts and observe the little voices in your mind. What
do your hear in your mind when you worry or fret about an upcoming exam?
What thoughts get you nervous? What are your worst fears? What thoughts
stop you in your tracks and sabotage you? What negative things are you
telling yourself? Write them down here.

Once you identify your negative self-statements and bring them
into the open, you can deal with them more rationally and effectively.
The next step is to challenge each of these negative, self-defeating

statements and reverse them to positive, self-enhancing statements. Examples of negative to positive shifts include the following:

Negative Self-Statements	Positive Reversals
I can't do it.	I know I can do it.
I feel overwhelmed.	I can do this one step at a time.
It's too difficult for me.	I have the aptitude to learn this.
I'm stupid.	I have good abilities.
Everyone else knows how to do this but me.	I am learning how to do this.
I'm not really smart.	I know I am capable.
I'm too nervous to concentrate.	I am remaining calm and relaxed even under pressure.
I'll flunk, so why try?	I am learning more each day; success is bound to follow.
I hate tests.	Tests are becoming easier for me.
Who cares about this test anyway?	This test is a positive challenge for me to show what I've learned.

EXERCISE 9-7 Negative to Positive Self-Statement Shifts

In the following spaces, create your own negative to positive shifts. List the negative thoughts causing you to be anxious before an exam and then shift them to positive, achievement-enhancing ones. Be sure to use only positive terms. Avoid all conditional statements ("should" and "could"), and write the positive statements in the present tense. Although you may not at the moment believe the positive statement is true for you, it is nevertheless necessary to state it: no "wills," "ifs," or future tense.

Negative Self-Statements	Positive Self-Statements
_____	_____
_____	_____
_____	_____
_____	_____

Negative Self-Statements	Positive Self-Statements
_____	_____
_____	_____
_____	_____
_____	_____

This process of changing negative to positive self-statements will help to demystify and disempower your negative self-talk, and eventually, it will fall by the wayside. Every time a negative thought or self-statement reappears, you should reverse it to a positive statement.

If, despite all efforts, some negative thoughts continue to plague you, a useful technique to overcome them is the thought-stoppage procedure.

EXERCISE 9-8 Thought-Stoppage Technique

Use this technique whenever a disturbing or distracting thought comes to mind. First, in your mind, you are to yell "STOP," and then say "calm." Then deliberately relax by taking several deep, comfortable breaths. The aim is to get at least a momentary break in the negative thought process, making it possible to achieve conscious control and regulation of self-defeating inner dialogue. Each time the negative or disturbing thought comes to mind, repeat this procedure: yell "STOP," say "calm," and then deliberately relax. Allow this technique to become automatic, and soon the negative thoughts will be completely eliminated.

EXERCISE 9-9 Positive Statements for Academic Success

To keep inner dialogue positive and reinforcing, repeatedly conjure up positive statements, and visualize them being true. In the following list, check off the positive statements you think would be most rewarding to you (check at least five). Be sure to add some of your own.

_____ I'm a good student.

_____ I'm learning more each day.

_____ I am capable.

_____ I have good abilities.

_____ I allow myself to relax while studying.

_____ My memory is improving each day.

_____ My mind is clear and alert.

_____ I see myself accomplishing my goals.

_____ I eliminate all distracting and disturbing thoughts.

_____ My comprehension is excellent.

_____ I am an intelligent, talented person.

_____ I have confidence in myself.

_____ I have good concentration.

_____ I'm remaining calm and relaxed even under stress.

_____ I'm handling my studies well.

_____ I am managing my workload.

_____ I'm getting better each day.

_____ I have studied well, and I know my material.

_____ I'm motivated.

_____ I can do it!

_____ I'm making it!

_____ I remain calm, relaxed, and alert on my math exam.

_____ I remain clear, capable, and confident on my math exam.

My positive statements:

1._____

2._____

3._____

4._____

5._____

6._____

I ask students to write positive statements such as the ones in Exercise 9-9 and post them on their mirror, on the refrigerator door, or in the place they usually study for exams. Soon magical things happen. Students report they feel better, they study better, and they're calmer during tests. Say and visualize positive statements and you, too, can create self-fulfilling prophecies. You will begin to feel better. Your attitude will improve, and your anxiety level will decrease. You will be able to function better and more efficiently. Always keep in mind whenever negative self-statements creep into your thoughts, either during studying or test taking, you must push them away, tell them to stop, and then deliberately relax. You must focus on relaxing, remaining positive, and refocusing your attention on your studies or the test at hand.

Relaxation

This brings us to the second part of my three-pronged approach for dealing with the psychological aspects of test anxiety: relaxation. One of the most important things you can do in the weeks, days, and even minutes before an exam is to practice calming and relaxation techniques. You should *not* study up to the last minute before an exam. Repeatedly, I've seen students who did this and then went into the exam and panicked or completely blanked out. Instead, in the last half-hour before the exam, I recommend you calm yourself down using the Calming Breath or Deep Abdominal Breathing techniques recorded on the CD accompanying this workbook and described in Chapter 3. By practicing slow breathing from deep within your lower lungs, you will gain control over your anxiety response and prevent it from accelerating and escalating to high levels. Remember, moderate, manageable anxiety is the key to successful performance.

EXERCISE 9-10 **Practice the Calming Techniques (See Chapter 3)**

Practice a deep, slow breathing pattern for a few minutes each day in the days before your exam and in the half-hour prior to your exam. Fill your lungs completely, feel your abdomen expand, hold the air a moment, and then slowly exhale.

"OK CLASS – READY FOR YOUR TESTS?"

Test-anxious students find it especially beneficial to keep to themselves before the exam. Don't get to the exam too early, and avoid talking with classmates, if possible. This is important for two reasons. First, it prevents you from picking up other students' anxiety or from becoming anxious because of something they might say. Second, the first recall after studying is usually the best. I've met students who reported someone asked them a question before the exam, and they responded correctly but then proceeded to get the problem wrong on the exam. So be a hermit! When you do arrive for the exam, find a comfortable seat and begin to focus on being calm, positive, and relaxed.

Maintain a calm, slow breathing pattern throughout the exam, while continuing to say positive self-statements. Push away all distracting or disturbing thoughts and focus on staying positive and calm. If you feel your anxiety increasing during the exam, stop and do Deep Abdominal Breathing. You are in control. By keeping your anxiety at a moderate level, you will be able to think clearly and do your best.

Visualization

The final part of my approach for dealing with the psychological aspects of test anxiety uses visualization. Here, while in a deep state of relaxation, you are to visualize yourself feeling confident, capable, relaxed, and successful while taking your math exam. It is important to practice the "Math Test Anxiety Reduction Visualization" (Exercise 9-11) each day for at least 5 days before the exam! The CD accompanying this workbook contains this visualization on track five. Listen to it often. It is particularly important to practice it the evening before and the morning of the exam. Listen in a peaceful, quiet, darkened place where you can be alone and undisturbed.

EXERCISE 9-11 A "Math Test Anxiety Reduction Visualization"

Begin by assuming a comfortable, relaxed position. Let your body settle down inside and, gently and slowly, close your eyes. It is easier to concentrate on relaxation with your eyes closed.

Let's begin by focusing on your breathing for a few moments. Take a deep and comfortable breath, filling your lungs completely . . . hold it a moment . . . and then . . . very, very slowly, let it out . . . slowly . . . feeling a wave of relaxation going from the top of your head all the way down to your toes. . . . Now I would like you to take another deep and comfortable breath, filling your lungs completely . . . hold it a moment . . . and then . . . once again let it out . . . very, very slowly, feeling another wave of relaxation going from the top of your head all the way down to your toes . . . slowly . . . feeling more and more relaxed.

Continue to breathe slowly, deeply, regularly. As you breathe in, concentrate first on filling the lower part of your lungs, then the middle part, and then the upper part. Feel your abdomen slowly expanding . . . now exhale very, very slowly . . . slowly, saying "relax" as you do so. Continue breathing slowly and regularly, filling your lungs completely and then so, so slowly exhaling, saying "relax," and *feeling* relaxed. Allow each breath to take you into a deeper and deeper state of relaxation. Continue alone for a few minutes. (Pause for 2 minutes.)

Feel your mind and body becoming more and more peaceful and serene. . . . And now, beginning at the top of your head, allow all the tension to leave your body, relaxing further and further. Let go of all the tension in your scalp, forehead, and the tiny muscles around your eyes. . . . Relax your jaws. . . . Allow your neck and shoulder muscles to go loose and limp. . . . Continue to breathe slowly and deeply, saying "relax" with every exhalation. . . . Feel all the tension drain out of your arms and hands. . . . Now allow your chest, abdomen, and back to relax. . . . Feel the relaxation flowing through and soothing each and every vertebra of your spinal column, from the top all the way down to the bottom. (Pause for 30 seconds.) Now feel the tension draining from your hips, down your legs, and out the bottoms of your feet. You are relaxing more and more with every breath you take. (Pause for 20 seconds.)

Now say to yourself and visualize the following (pause for 5 to 8 seconds between statements):

> I see myself accomplishing my goals.
> I can eliminate all distracting and disturbing thoughts.
> I have confidence in myself.
> I am an intelligent, talented person.
> My memory is improving each day.
> I see myself relaxed and calm on my math exam.
> I see myself relaxed, calm, alert, and confident on my math exam.
> I am relaxed, calm, capable, and confident on my exam.
> I see my math exam going well for me.
> My mind is calm, clear, and alert during my math exam.
> I see myself succeeding in math.
> I now say to myself, over and over, my own special positive affirmations
>   and visualize them being true for me.
> (Pause for 1 minute.)

Next I would like you to imagine that you have just finished all your studying and reviewing for a math exam tomorrow. . . . You are feeling good. You have studied well, and you are confident that you will do well on the exam tomorrow. . . . You get ready for bed and you snuggle down under the blankets, but before going off to sleep, you practice relaxation. You slowly and deeply breathe in and out . . . saying "relax" with every exhalation and visualizing the exam going well tomorrow. . . . You soon fall off to sleep knowing you will have a good night's sleep. . . .

Now imagine yourself awakening the next morning. You are feeling good. You have had a good night's sleep, and you feel refreshed, clear, and alert. You eat a nutritious breakfast of low-fat protein to help sustain your alertness. You're feeling so good. You recall all the studying and reviewing your have done in the past weeks for today's exam, and you *know* that you know the material. You have organized yourself well, reviewed often, and you have confidence that you will do well on the exam today. . . . Again you spend a few minutes relaxing and visualizing the exam going well today. . . .

Now imagine that you are on your way to school. What are your surroundings like? Is the sun shining? What color is the sky? How does the air feel against your cheek? Now see yourself walking toward the building where your test will be given. Notice the colors, the sounds, the smells, the shadows. You are now entering the building and slowly approaching the room. You are not too early or too late for the exam. You have scheduled your time well to arrive just in time. You are like a hermit before the exam; you avoid any contacts or discussion about the test with anyone. You keep yourself in a peaceful, calm state. You slowly enter the room. You find a seat.

As you sit quietly in your chair waiting for the exam to be passed out, you close your eyes or focus them on a point of light in the distance. You tune out the other people in the room and turn inward for a few moments. You breathe slowly and deeply from your lower lungs, and you take yourself through the relaxation process you have practiced the last few days, saying "relax" with every exhalation. As you do this, you feel all the tension leave your body, dissipating into nothingness. A sense of well-being comes over you. As you sit there, you recall the studying and reviewing you did for this exam; you see your lecture notes, your textbook notes, the many problems you've worked out. You see the formulas, the important concepts, and the solutions to difficult problems. You recall all the time and effort you put into reviewing all the information, and you *know* that you know your work. You are feeling capable and confident. You are calm and relaxed, yet clear and alert. (Pause for 15 seconds.)

Now imagine that you are given the exam. You read the directions carefully, underlining all significant words so that you won't overlook them or misinterpret them. You skim through the whole test, noticing the ideas covered, the types of problems, and the point values. You smile to yourself because you recognize all that is being asked of you, and you feel good about your preparation. (Pause for 10 seconds.)

You proceed confidently through the exam, beginning with the questions that appeal to you. You don't necessarily begin with the first question if you don't want to. You may choose the ones that seem easiest or most fun. By doing these first, your confidence grows. You feel good. You feel alert, assured, and calm. (Pause for 20 seconds.) You continue through the exam, pacing yourself well. You recall your notes, handouts, homework assignments, and class discussions. You proceed confidently through the exam. When you come to difficult questions that you are not sure how to answer, you leave them for a while and simply go to easier ones. As your confidence builds, you soon return to these difficult questions, and the answers come to you much more readily. With some of the difficult questions, you simply sit back, take a deep breath, exhale slowly, and say to yourself, "relax." You soon find that the answers to these questions come to you much more readily and quickly. You have studied all the information. You know you know the work. You are very capable. By staying calm and relaxed, you are able to function at your highest level on the exam. You are feeling calm, relaxed, confident, and capable. (Pause for 15 seconds.)

Now I would like you to imagine that you have completed the exam. You spend a few minutes checking it for errors, omissions, and reasonableness of the solutions. (Pause for 15 seconds.) You are feeling relaxed and calm and in no hurry to leave the room until you have finished checking your paper. . . . Now imagine that you hand in the exam paper. You leave the room. Once in the fresh air, you take a deep breath and exhale slowly. . . . You feel refreshed and happy. The exam has gone well, and you have done the best job you could. You are feeling wonderful. You have remained calm, alert, confident, and capable throughout the exam. The exam has gone well for you, and you know that you will remain calm and relaxed for exams in the future. . . . Be sure to give yourself a special treat later in the day for handling the exam so well.

Now I would like you to slowly and gently bring yourself back to this room, carrying with you the pleasant feelings and thoughts that you experienced during this exercise.

Performance Health Tips

It has been said the body is the temple of the soul. I have often found test-anxious students to be poor overseers of the golden temples in which they reside. They eat poorly, sleep minimally, and exercise rarely. Battling fatigue and nervous exhaustion with little or no fuel or resources, they easily fall prey to their nemesis—test anxiety.

Many students get far fewer than the recommended 7 to 8 hours of sleep each night. Fatigue and exhaustion reduce efficiency and cause poor memory and recall. Time for relaxation and meditation also may be missing from a student's busy schedule. Twenty minutes of reverie or silent contemplation can often make up for a shortened night's sleep.

Students also have a tendency to stop exercising in the days or weeks before a test. A regular program of moderate exercise (for example, walking or biking for 20 minutes a day) can greatly reduce stress, as well as increase alertness, clear thinking, and energy level.

For many students, caffeinated beverages such as coffee or soda are the culprit. In an attempt to be more alert for a test, students may decide to drink more caffeine than they usually do or drink caffeine when they are not normally caffeine drinkers. Unfortunately, excessive caffeine or caffeine taken by people who are not accustomed to it can only make them jittery, "hyper," or shaky for the test. If they were already anxious, this added shakiness may cause them to panic!

Poor nutrition, such as an unbalanced diet or eating high-caloric, fatty meals, can negatively affect a student's thinking and problem-solving abilities.

Judith Wurtman's *Managing Your Mind and Mood through Food* contains an excellent discussion of the effects the foods we eat have on our brain's ability to process information. Tyrosine, for example, an amino acid found in protein, is the principal ingredient in two neurotransmitters. These neurotransmitters, dopamine and norepinephrine, produce alertness, clarity of mind, motivation, and drive. Research shows that with an adequate supply of these two neurotransmitters, people tend to think more quickly, react more rapidly, feel more attentive, and experience greater mental endurance and energy.

Wurtman's studies show eating only 3 or 4 ounces of low-fat protein foods, either alone or with carbohydrates, makes tyrosine available to the brain if the brain has used its current supply and needs to replace it. This process has an energizing effect, increasing alertness, accuracy, and motivation. The brain responds more quickly and is prepared for mental challenges.

The best foods to choose for heightening brain power before an exam include 3 or 4 ounces of any one of the following:

fish
chicken without the skin
veal
very lean beef with all the fat trimmed
nonfat or low-fat yogurt
low-fat cottage cheese

tofu
lentils, dried peas, or beans

Avoid large meals and high-fat foods before tests because they tend to dull the mind and slow mental processes. They produce lethargy and drowsiness. You must avoid foods such as lamb, pork and pork products, luncheon meats, hard cheeses, fatty beef, whole milk regular yogurt (as opposed to nonfat or low-fat), butter, mayonnaise, fried foods, creamed soups, and rich gravies and sauces.

In addition to tyrosine, another amino acid, tryptophane, is found naturally in food and significantly affects our brain's ability to function. I am referring to tryptophane in its *natural* form, *not* in a pill. Naturally occurring tryptophane is the principal ingredient in the neurotransmitter serotonin.

Serotonin has a calming effect on the mind. It reduces feelings of tension and stress. Concentration and focus improve. Distractions are more easily eliminated. Reaction time may be slowed. And if you eat serotonin-producing foods in the evening or before bedtime, they may induce drowsiness or sleep.

Eating only 1 or 1.5 ounces of carbohydrate alone, without protein, increases your brain's level of serotonin. You will experience a calming, relaxing, more focused effect. Your powers of concentration will increase, and your feelings of anxiety and frustration will ease. People who are 20% over their recommended weight or premenstrual women may require 2 or 2.5 ounces of carbohydrates to get this effect.

The best foods to choose for producing a calming, focused effect while studying include:

cereals	crackers	popcorn
muffins	pasta	potatoes
bagels	rice	bread
pancakes	barley	corn

Here is a summary of recommended health guidelines:

1. Make it a rule to get 7 to 8 hours of sleep regularly, especially before tests.
2. Allow time each day to meditate or relax. This is particularly important if you had little sleep or experienced insomnia the night before a test.

3. Participate in a program of moderate exercise for at least 20 minutes each day or every other day. Don't stop exercising around test time.

4. Eating for high mental energy requires you eat small, low-caloric, low-fat meals and you get proteins into your system before, or along with, carbohydrates. So do not begin your meal with carbohydrates, but with protein. Remember: You need only 3 or 4 ounces of protein. High amounts of protein may actually have a reverse effect on some people. Ideal foods include fish, chicken, veal, nonfat yogurt, tofu, lentils, and peas.

5. Time your high-energy meal to occur about 1 to 2 hours before you must meet the mental challenge of taking a test.

6. To help you get adequate sleep before the test, eat high-carbohydrate, low-protein (or even better, no protein) foods for dinner or in the evening to induce drowsiness and sleep.

7. If at any time while studying you feel mentally frustrated, scattered, and can't seem to settle down, eat small amounts of carbohydrate to help calm you and focus your concentration. Although carbohydrates in the form of sugar also have this effect, be sure to eat no more than 1 or 2 ounces. Too much sugar causes negative emotional reactions in some people. Ideal carbohydrates to help settle you down are popcorn, bread, crackers, muffins, bagels, potatoes, rice, and pasta.

8. Don't overeat or eat foods with high fat content. These will only make you lethargic and dull.

9. Avoid excessive use of caffeinated beverages, such as coffee or soda. If you don't normally drink caffeine, don't start now, particularly not before a test.

Be Your Own Exam Coach: Put It All Together

As a coach, you can boost your self-confidence, give yourself pep talks, keep yourself focused on the task at hand, push away distractions, and set realistic, achievement-oriented exam goals.

Too many math-anxious students set weak or ineffective exam goals: They do not want to do badly or to fail or appear dumb. Avoidance goals such as these lead a student in the opposite direction from success. They may lead directly to fears, worries, negative thinking, and worst of all, increased anxiety and stress.

If you've set avoidance goals in the past, why not change your approach? What if you coached yourself to set exam goals realistically related to what you could strive for and achieve? Here are examples of such goals:

1. I want to achieve at least 10 points higher than I did on my last exam.
2. I want to demonstrate to myself that I have mastered these math skills.
3. I want to remain calm and relaxed on this exam and be able to think clearly.
4. I want to tackle the easiest questions first, then the next easiest, and so on, until I can complete all the questions on the exam.

EXERCISE 9-12   Collaborative Learning: The Best Coaching Practices

With a small group of classmates, discuss what you believe would be the best way for you to coach yourself for exams. In your discussion, answer some of the questions that follow. When you are finished, jot down coaching ideas you think would help you on your next exam.

- What positive, achievement-oriented goals can you set for your next exam?
- What strategies can help you stay focused on the task?
- What techniques are helpful to keep distractions and disturbing thoughts away?
- How can you boost your confidence and keep your spirits up during the exam?

Reflections:

_____

_____

_____

_____

EXERCISE 9-13 Math Test Anxiety Reduction Checklist

Use this checklist *before* your next exam.

_____ I've reviewed and worked out lots of problems so I know my material out of context.

_____ I know the format and content of my upcoming math exam.

_____ I know how many questions will be on my exam and its duration.

_____ I've given myself several practice exams.

_____ On practice exams, I've noted areas of difficulty so I can strengthen them.

_____ I've analyzed my past pattern of typical errors so I can be alert to them on my exam.

_____ I've gotten 7 to 8 hours of sleep in the days prior to the exam.

_____ I've kept up a regular program of moderate exercise.

_____ I've practiced relaxation exercises along with positive visualization in the days and the half-hour before the exam.

_____ I've eaten a small meal of low-fat protein 1 to 2 hours before the exam and avoided too much caffeine.

_____ I'll arrive at the exam on time and avoid talking with others.

_____ I will coach myself throughout the exam using positive coaching practices.

_____ Throughout the exam, I'll remain calm, relaxed, and positive, checking my breathing often.

_____ I will say positive self-statements to myself and push away all disturbing or distracting thoughts.

_____ I will write out all my formulas and key ideas on the top corner of my exam sheet before beginning the test.

_____ I'll quickly read through the exam, note point values, and schedule my time accordingly.

_____ I'll proceed comfortably throughout the exam, working first on the problems that come most easily to me.

_____ I'll carefully read the directions to all problems and circle significant words to avoid misinterpretation.

_____ After finishing the exam, I'll check my answers, proofread for omissions, and check for my typical errors.

_____ I'll leave and reward myself for a job well done!

_____ I have set a positive achievement-oriented goal for this next exam, and it is:

_____

_____

_____

After Your Next Exam

After you complete your exam, reflect on what you learned about yourself and your test-taking skills. Analyze what went well for you and what strategies worked best. Determine which strategies need to be modified and which ones will help you on future exams. Here are important questions to ask yourself:

What worked best for me on this exam?

What study techniques really paid off on this exam?

Was I adequately prepared? If not, what should I do differently next time?

What techniques helped me remain calm and relaxed?

On what do I still need to work?

What difficulties came up and how can I prevent them from interfering next time?

What helped me stay focused and kept distracting or disturbing thoughts away?

What things should I do differently when I study for the next exam?

What better ways can I use my time and energy on the next exam?

What should I remember to repeat next time?

Did I take good care of myself physically?

Do I need more sleep before my next exam?

Did I achieve the exam goal I set for myself?

Reflections: In this space, write your reflections.

_____

_____

_____

_____

Successful math students are able to function at an optimal level on exams. Exercise 9-14 asks you to become aware of some of the characteristics of optimal test performance. Use this checklist to see how many of these statements are true for you. Why not set a goal for yourself to be able to check more and more of these statements after future exams.

EXERCISE 9-14 Characteristics of Optimal Test Performance: A Self-Assessment

In the first column are characteristics of optimal performance on tests. In the second column, check whether you experienced any of these characteristics on this exam. Each time you take an exam, see if you can experience more of the characteristics listed in this self-assessment.

Characteristics of Optimal Test Performance	Check
I was mentally alert	_____
I was able to think clearly	_____
I pushed distracting thoughts out of my mind	_____
I felt good about my problem-solving skills	_____
I felt enthusiastic about thinking during the test	_____
I did some creative thinking	_____
My perceptions were sharp	_____
I was very motivated to do well	_____
I felt that I exercised good judgment	_____
I used my time and energy efficiently	_____
I was well organized	_____

I was able to work quickly   _____

I remained positive throughout the exam   _____

I was calm under pressure   _____

I was able to remember and recall my material   _____

I felt a sense of control   _____

I checked over my work and made minimal careless errors   _____

I was self-confident   _____

Summary

Don't allow test anxiety ever again to detract from your math achievement. What once was an overwhelming and fear-producing event can now be handled in a constructive, positive, growth-enhancing manner. This section of our road map focused on methods for markedly reducing math test anxiety. I've stressed good nutritional habits, adequate sleep, relaxation and exercise, and effective test preparation and test-taking strategies. In particular, I highlighted ways to deal with psychological factors through a special three-pronged approach. This approach includes a combination of shifts from negative to positive self-talk, calming techniques, and the utilization of my specially designed "Math Test Anxiety Reduction Visualization." I concluded this chapter with an exercise to put it all together, the Math Test Anxiety Reduction Checklist and a self-assessment of optimal performance. I encouraged you to be your own coach, to set positive exam goals, and to reflect on what worked for you on your test.

CHAPTER

10

Be a Successful Problem Solver

In Greek mythology, Sisyphus, the greedy King of Corinth, incurred the wrath of the supreme deity, Zeus. Sisyphus was punished by being forced to toil eternally, pushing a huge boulder up a steep hill. Each night, he'd fall asleep, exhausted, only to have the boulder roll back down again. Each day, this endless, toilsome, useless process was repeated.

Like Sisyphus, we often spend our lives in routine tasks made difficult only by our perceived individual limitations. We set up self-imposed boundaries. We limit our vision. We do things only one way, without experimenting with alternative approaches.

Barriers to accomplishments are unproductive and must not be maintained. We must challenge the status quo. We must open our minds to new ways of thinking. How wonderful to choose to be part of the solution, not the problem.

To some students, solving math equations and handling word problems can loom as menacingly as Sisyphus's boulder in the Greek myth. But it's just that—only a myth. So stand back, take a deep breath, and realize that information, insight, and understanding can soften the harsh lines of resignation we draw in the sand. Comprehension breeds success, but there are tricks to it. Academic success hinges on learning the tricks.

A physicist friend once shared her secret for academic success. In college, she realized her teachers used thinking approaches unique to their disciplines. For example, her math teacher thought quite differently from her biology or psychology teachers. She found if she could train herself to think like a chemist, or a sociologist, or a mathematician, or for that matter, like any other subject area specialist, she was successful in learning the subject material.

How do mathematicians think? One math instructor told me, "Math is to the mind as weight training is to the body." What an interesting concept. Math does strengthen our thinking, just as a well-planned exercise program strengthens our muscles. Since the heart of math is problem solving, mathematicians are actually doing "mind building" when they're solving problems. In this chapter, we work on taking the mystery out of how mathematicians think. We also examine approaches to problem solving and word problems.

Mathematicians train their minds to think logically. In this type of thinking, when given a problem, the person proceeds step by step and uses specific rules to draw a conclusion or derive a solution. This is not to say problems can't be solved in other ways, for they often are. Some students intuitively solve equations or problems. They are certain they know the answer but cannot explain how they got it. Others use their own personal approaches to problem solving and have been successful. There are still others who look at how similar problems are solved, and through comparison and association, they figure out the correct solution. These are all fine ways to solve problems, though none is a fail-safe method for consistent success.

If you are one of these people, in addition to using your own problem-solving strategies, I would like to encourage you also to practice how mathematicians think. By using their disciplined, logical thinking approach, you will have more resources to successfully tackle and solve more and more math problems. Like a gold medal Olympic gymnast, you can go for the *gold*.

When thinking like a mathematician, you will also be enhancing your left- and right-brain thinking. Our brains have two cortical hemispheres. The left hemisphere is superior in language, math, judging time and rhythm, and at ordering complex movements. This hemisphere can analyze things and break information into smaller and smaller bits. It can process information sequentially, one bit at a time, one after the other. The mathematician's logical approach fits into left-brain thinking. On the other hand, the right hemisphere is superior in nonverbal skills such as perceptual and spatial skills, visualization, drawing pictures, putting puzzles together, simple language comprehension, and recognizing faces, patterns, and melodies.

The right brain tends to process material simultaneously, in a multidimensional fashion, as if all aspects of the situation are seen at once. Intuitively knowing how to solve a problem fits into right-brain thinking.

Most people do not rely on only one hemisphere. Brain studies show there is an interplay between the left and right hemispheres in almost all human activities. In fact, the most creative thinkers are engaging both left- and right-brain thinking on a continuous basis. However, there are some people who tend to prefer either left- or right-brain thinking. These individuals may limit their career or college major choices to those highlighting their thinking preference. They may choose certain activities more often than others. They would approach problem solving and decision making differently from those who use both brain hemispheres more equally. Exercise 10-1 will help you see if you are more right-brained or left-brained when solving problems.

EXERCISE 10-1 Are You a Left-Brain or Right-Brain Problem Solver?

As you know, most people use both sides of their brains for dealing with virtually all activities. In fact, the best problem solvers are constantly using both sides of their brains. However, some people prefer a thinking style favoring their right or their left brain. You may be one of these people. To determine this, carefully read the following statements. Give yourself three points if the item usually applies to you, two points if it sometimes applies, one point if it rarely applies, and zero if it never applies.

A. Are You a Left-Brain Thinker?

_____ 1. Before making a decision, I like to compare and contrast many ideas.

_____ 2. I would prefer settling a disagreement based on what is fair and reasonable.

_____ 3. I like working within a structured format where things follow a logical sequence.

_____ 4. I prefer to use my experience and knowledge of the facts to make a decision.

_____ 5. When solving problems, I prefer screening, selecting, and supporting new ideas.

_____ 6. I usually remember diagrams or schematic drawings better than people's faces and names.

_____ 7. I prefer improving and refining alternative options.

_____ 8. Rather than exploring new ideas, I prefer coming to closure and making effective decisions.

_____ 9. I enjoy analyzing and developing new ideas to take effective action.

_____ 10. I really prefer completing everything I start.

_____ TOTAL SCORE

B. Are You a Right-Brain Thinker?

_____ 1. When making a decision, I prefer thinking up many possibilities and alternatives rather than evaluating these ideas.

_____ 2. I prefer settling a disagreement based on what makes people happy rather than on what is reasonable or fair.

_____ 3. When working on a project, I prefer having complete flexibility in an unstructured environment.

_____ 4. I prefer using my intuitive sense and my feelings when making a decision.

_____ 5. I prefer thinking up or brainstorming new and unusual ideas.

_____ 6. I usually remember faces and people's names more easily than diagrams or schematic drawings.

_____ 7. I enjoy thinking and experiencing things in different ways, using alternative points of view.

_____ 8. I enjoy exploring new ideas and possibilities in depth.

_____ 9. I enjoy making meaningful new connections between ideas.

_____ 10. Completing a project isn't as important as enjoying the process.

_____ TOTAL SCORE

Analysis of Results:
A score of 22 or above on section A, with a score of 18 or below on section B, indicates you have a strong preference for left-brain thinking. On the other hand, if you have a score of 22 or above on section B, with a score of 18 or below on section A, you probably have a strong preference for right-brain thinking. If your scores on sections A and B are similar (within 1 or 2 points of each other), you are not showing a preference for using one side of your brain over the other. Scores of 22 or above on both sections show a strong preference for using both sides of your brain when solving problems.

When you jump out of an airplane with a parachute on your back, your life is on the line. Your survival depends on following a well-practiced sequence of known steps. When you rock climb on a sheer 100-foot wall, your life is on the line. Your survival depends on following a well-planned sequence of steps. When you scuba dive for sunken deep-sea treasure, your life is on the line. Your survival depends on following a well-rehearsed sequence of steps. Although these examples may seem extreme, they do illustrate an interesting and little discussed point. Risk taking and the mathematician's logical problem-solving approach share many features. They both rely on left-brain function.

Psychologist Roger Drake and others found that people have an increased willingness to make risky decisions when their left hemisphere is more active than their right. Research shows the right hemisphere is associated with a more cautious cognitive style. So if your preference is right-brain thinking, you may be more hesitant and wary when approaching math problems. You may be less likely to jump in and start working with the numbers and figuring out solutions. Rather than risk being wrong, you might never even try. Is this you? If so, I would like to challenge you to work on stimulating your left-brain thinking. Work on practicing the mathematicians' logical problem-solving approach. This will help you take more risks in solving math problems, and with this increase in your risk taking, you will certainly add to your potential for math success. Exercise 10-2 provides you with some suggestions and experience for practicing left-brain logical sequencing skills.

EXERCISE 10-2 Practice Logical Sequential Thinking

A. Games You Can Play

Many games test your power to think logically and follow a sequence of steps until you arrive at a conclusion. You might enjoy playing these games.

1. Master Mind
2. Clue
3. Chess
4. Backgammon
5. Contract Bridge

B. Create a Positive Force Field

Use this technique to deal with difficult situations in your life.

1. Describe the situation you would like to solve.
2. How would you prefer this situation to be resolved? Make a list describing the best possible conclusions.
3. Next to this list indicate what positive forces or strengths you have going for you that can push you toward this positive conclusion.
4. Now ask yourself: What is the worst possible resolution to this difficult situation? Make a list of the characteristics of this negative resolution.
5. Next to this list indicate the negative forces pushing you in that direction.
6. Compare the positive forces listed in step 3 with the negative ones in step 5. Describe at least three ways you can strengthen or add to the positive forces and eliminate or weaken the negative ones. When you can reduce or wipe out the negative forces, you will be able to achieve your most desired results. Good luck!

C. Use a Top-Down Programming Approach

Level 1. Describe a difficult situation you would like to solve.

Level 2. List three or four possible ways to approach the problem.

Level 3. For each of the approaches listed, identify at least two steps you'll need to take to implement the approach.

Level 4. For each step listed in level 3, identify what further steps need to be taken. Continue this process, listing the smaller and smaller steps needed.

At the top of page 178 is a diagram of one possible top-down programming approach.

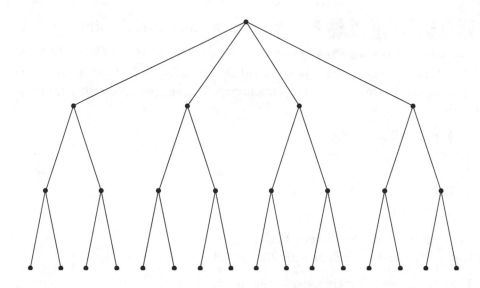

Polya's Problem-Solving Tricks

In mathematician George Polya's famous book, *How to Solve It* (Princeton University Press, 1971), he published his classic problem-solving model, along with a host of suggestions, advice, rules of thumb, and strategies for solving problems. His problem-solving model has four steps:

1. Understanding the problem.
2. Devising a plan.
3. Carrying out the plan.
4. Looking back.

This model will help you follow a logical thinking approach, starting with some given information and proceeding step by step until you reach a reasonable conclusion.

In the first step, ask yourself to restate the problem in your own words. What is this problem really asking you to do? What are the unknowns involved? What information does the problem give you and what is missing?

In Polya's second step, you look for patterns. Ask yourself: Is this problem related to previous problems you have solved, preferably simpler ones? Can the same technique used on previous problems be used here? If you can't find a similar related problem, is it possible to restate the present problem? Can you draw a chart, table, diagram, or model to illustrate what is being sought? Can you work backward? You may wish to "guess and test."

In the second step, if you can't figure out how to solve the problem, can you divide the problem into smaller parts and identify subgoals to solve? This is also known as "divide and conquer." But no matter what, don't get discouraged, because even Polya says, in addition to formerly acquired knowledge, good mental habits, and concentration on purpose, you may need "good luck." This part of Polya's model certainly brings in some right-brain thinking.

In Polya's third step, you are to carry out a plan. Here, you should perform the necessary computations. Do all the geometric or algebraic operations you identified in step two. Check each step of your plan as you proceed, either using a formal proof or your own "intuitive" approach. What's neat about Polya's guidelines is his integration of intuition with logic. Again, you can see the importance of incorporating right- and left-brain thinking.

In the final step, you can check the result of your problem and interpret the solution in terms of the original problem. Ask yourself: Is there another, more direct solution?

After your solution is found, Polya urges you not to shut your book or look for something else to do. Try to determine other related or more general problems for which your results or methods would work. He says, "By looking back at the completed solution, by reconsidering and reexamining the result and the path that led to it," you can consolidate as well as acquire well-ordered, ready-to-use knowledge and further develop your ability to solve problems.

Polya's problem-solving approach is certainly very straightforward and makes good sense. Now let's apply it to a word problem presented earlier in this book (see Exercise 4-3):

> A business executive left her car in an all-night parking garage for three and a half days. The garage charges either an hourly rate of $1.25 or a daily rate of $24 for whole or partial days. Is she better off paying the hourly or the daily rate?

Step 1: Understanding the Problem

What is this problem really asking you to do? What are the unknowns involved? What information does the problem give you and what is missing?

This problem is asking us to determine which rate would be cheaper for the executive to pay. The problem tells us the hourly rate is $1.25, the daily rate is $24, and our business executive stayed 3 and a half days, which is equivalent to 24 (hours) times 3 (days) plus 12 (the hours in half a day), or a total of 84 hours. We must find the cost for 84 hours using the hourly rate of $1.25. Since the daily rate will be charged for whole days, the cost of 3 and a half days would be the same as 4 whole days.

Step 2: Devising a Plan

Can you draw a chart, table, diagram, or model to illustrate what is being sought? Can you divide the problem into smaller parts and identify subgoals to solve?

This problem can be solved by identifying the subgoal to solve—that is, multiplying out the hourly and daily rates. A table would be very helpful for easy comparison. The table would look like this:

Hourly Rate	Daily Rate
84 hours (same as 3 days, 12 hours) × $1.25	4 days (actually 3 days, 12 hours) × $24
?	?

Step 3: Carrying Out the Plan

You should perform the necessary computations.

Hourly Rate	Daily Rate
84 hours (same as 3 days, 12 hours) × $1.25	4 days (actually 3 days, 12 hours) × $24
$105	$96

From this calculation, it is evident that the daily rate would be cheaper than the hourly rate.

Step 4: Looking Back

Now look back at the completed solution; reconsider and reexamine the result and the path that led to it. Check the results of your problem and interpret the solution in terms of the original problem. Is there another, more direct solution? Can you determine other related or more general problems for which your results or methods would work?

From this problem, we can see that setting up a table is an effective problem-solving method for comparing two or more price calculations. The table shows us that the daily rate, totaling $96, is the better buy. If combining both rates were a choice, we would have come up with an even lower price. In other words, paying the daily rate for 3 days and then the hourly rate for the 12 hours would have been $72 plus $15, totaling only $87.

Now that we've applied Polya's problem-solving model to a word problem, let's look at other tricks for tackling word problems.

Tackle Word Problems: Develop Your Own 12-Point System

I've encountered many students who find word problems difficult. When reading the problem, their minds become overwhelmed, they short-circuit, and shut down. At this point, nothing more goes in or out. They feel hopeless and close their books. If this ever happens to you, help is on the way.

I want to share with you some secrets I've discovered about word problems. First, it is important to realize all word problems are just made-up situations to challenge your thinking. To illustrate this point, I made up the two word problems in Exercise 4-3 in this textbook, and I am a psychologist and not a mathematician. So, you, too, can make up word problems.

The second secret is that words in a word problem can be changed into an equation. Your job is to figure out how to translate the words into an equation or equations that you have already learned in your class.

A third secret is that, usually at the end of the problem, you will discover the question you need to solve. When you are ready to set up your equation, you can let x stand for your unknown. If there are two equations involved, let y stand for the second unknown. Some mathematicians even try putting in actual number values for the unknowns to make the word problem more meaningful or realistic. This is a good trick to try.

The fourth secret: You can plug into your equation what you know from the word problem—that is, any relative numbers given in the sentences. Then you solve the equation for the unknown(s). This will be the solution to the word problem.

The last and most important secret is that, even for mathematicians, word problems take work. Mathematicians don't know how to solve problems on the first reading unless they've done the same problems before. It takes practice, organization, and a system. Of course, they may seem to solve them more quickly than you, but it is because they already have established an organization, a system, and a plan of attack. Many mathematicians tell me they often have to read word problems repeatedly before they know how to proceed. So don't expect the answers to word problems to jump out at you. Read the problem over and over again, set your plan of attack, and work.

My goal in this section is to give you an organization and a 12-step system to deal with word problems. This system works!

1. How can I restate the problem in my own words?
2. What is the problem asking me to do; that is, what question(s) must I answer?
3. What type of problem is it? For example, is it a money, age, investment, work, geometry, mixture, or a simple number story problem?
4. Can I draw a picture, chart, diagram, model, or table to illustrate the problem? (This will serve as a trigger and unblock your thinking. It helps you to "see" what's happening.)
5. Can I make a list of what the problem gives me (known variables) and what I will have to find (unknown variable[s]) and then put these into a table? (Usually, but not always, any given number in the problem will be a variable you can use.)
6. What equations have I learned that might apply here?
7. How are the pieces of the problem related to the equation(s) I have to solve?

8. Can I plug into the equation(s) the known and unknown variables I have in my table (see step 5)?
9. Do all the units I have plugged into my equation agree? If not, can I make them agree?
10. Can I state in words what I am going to do to solve this problem?
11. Now can I complete the solution?
12. If I plug my answer back into the problem, does my solution make sense? Is it logical?

Here are some helpful tips on how to translate the words in a word problem into parts of an equation:

- "is," "will be," or "was" becomes the equal sign in your equation
- "increased" or "more than" means addition
- "decreased" or "less than" means subtraction
- "twice as much" means 2 ×
- "half as much" means divide by 2
- "of" means multiply
- "square of" means a number multiplied by itself

Now let's try this system in a word problem I made up, and I hope we can have fun with it. But before beginning the problem, I want you to close your eyes and spend a few moments relaxing. Breathe from your abdomen using a slow, low breathing pattern. After you attain a calm state, open your eyes. You are ready to start.

Now slowly read through the word problem, while still keeping your breathing slow and regular. You may have to read the problem a piece at a time or perhaps you will have to read it two, three, five, or ten times before you can proceed. Then write down the answers to as many of the following questions as you can.

My Word Problem: One semester, I taught 97 students in three different math anxiety reduction classes. In class one, I had 8 students more than in class two. In class three, I had 7 students fewer than in class two. How many students were in each of the three classes?

Step 1. How can I restate the problem in my own words? "I taught three classes and the first and third classes are compared to the enrollment in the second class." Now, that was easy.

Step 2. What is this problem asking me to do; that is, what question(s) must I answer? The question is: "What's the enrollment in each class?"

Step 3. What type of problem is it? I'd say this is a simple number or algebraic story problem.

Step 4. Can I draw a picture? Well, yes. I'll let x stand for the number of students in class two.

BOX A		BOX B		BOX C		BOX D
Class two has x students	+	Class one has x plus 8 students	+	Class three has x minus 7 students	=	Total students = 97

Step 5. Can I make a list of what the problem gives me (known variables) and what I will have to find (unknown variable[s]) and then put these into a table?

	Knowns	Unknowns
Class one	$x + 8$ students	
Class two		x students
Class three	$x - 7$ students	
Total students	97	

Step 6. What equations have I learned that might apply here? Simple algebraic formulas with x plus or minus a number would work here.

Step 7. How are the pieces of the problem related to the equation(s) I have to solve? All parts of the problem can be easily dealt with in the equations.

Step 8. Can I plug into the equation(s) the known and unknown variables I have in my table? Answer: Yes.

$$x \quad + \quad x + 8 \quad + \quad x - 7 \quad = 97$$

Step 9. Do all the units I have plugged into my equation agree? Yes, they do. The units all refer to the number of students.

Step 10. Can I state in words what I am going to do to solve this problem? If I find the number of students in class two, I will know how many students are in classes one and three also.

Step 11. Now can I complete the solution?

$x + x + 8 + x - 7 = 97$	
$3x + 1 = 97$	I collect the like terms, which were 3 xs, and I subtract 7 from 8.
$3x + 1\,(-1) = 97 - 1$	I subtract the same number from each side.
$3x = 96$	
$\dfrac{3x}{3} = \dfrac{96}{3}$	I divided each side by the same number.
$x = 32$	Now we know that class two had 32 students.
$x + 8 = 40$	Now we know that class one had 40 students.
$x - 7 = 25$	Now we know that class three had 25 students.

Step 12. If I plug my answer back into the problem, does my solution make sense? Is it logical? Well, we can see that class one was 8 students more than class two and class three was 7 students less than class two. Best yet, all three classes add up to 97 students. So, yes, the answers are logical and make sense.

Identify the Exotic Animals and Deal with Algebra Problems

For some students, algebra is a stumbling block in their road to success. For them, algebra is too abstract, scary, meaningless, or intangible. The thought of dealing with a bunch of xs and ys seems overwhelming. Their eyes glaze over, and their minds go to sleep. They often ask, "Why study algebra?"

I have interviewed a number of students who have come to love and appreciate algebra. And I asked them just this question, "Why study algebra?" Repeatedly, they tell me the problem-solving skills learned in algebra help them deal with everyday life issues. They emphasize the benefits gained from algebra's logical reasoning, critical thinking skills, and a systematic, structured approach to tackling problems. Whether these students are budgeting for a vacation, figuring out a paycheck, or purchasing a house, they use algebra either directly or indirectly.

In this section, I review some helpful tricks for working on algebra problems. Although each problem you deal with will be different, by coming up with a problem-solving approach that works for you, you will find algebraic equations much easier to tackle. I would like to emphasize the most significant thing you can do to be successful in algebra is to practice effective problem-solving techniques and not worry about getting the correct answer. Your major goal is to understand what is being asked and to see the inherent logic in the question.

One of my favorite math teachers suggests you make algebra into a game. In this game, you make believe you are in a zoo and you are looking at a variety of exotic animals. Now, in this zoo, your challenge is to identify all the strange and unusual creatures before you. The same thing is true in the game of algebra. When you first look at a problem, ask yourself: What type of problem is this? Is it a linear equation? Is this a quadratic equation? Whenever you see an algebraic equation, immediately translate the equation into words. In other words, identify the "exotic animals." This game will take the mystery out of what you are looking at. It will make the abstract equation more concrete and meaningful to you.

Here are some tips you may wish to use after you have identified the type of algebra problem you are working with.

1. *Write down each step.* Proceed slowly from one step to the next. This will give you a sense of structure and a path to follow.
2. *Organize your work.* By arranging your work and following an order, you are more likely to successfully proceed from one step to the next in solving your problem.
3. *Put all your equal signs under one another.* In other words, line them up. This will reduce any confusion and help you clearly see each step of your solution.

4. *Be fair!* Since an equation is a statement that two quantities or expressions are equal, make sure you always balance both sides of the equation. A wonderful math tutor I know always tells his students to "Be Fair!" That is, whatever you do to one side, do to the other side of your equation.

   You can also think of your equation as a perfectly balanced rowboat. If you should put a large beer keg at one end of the boat without adding something of equal weight to the other end, your boat will sink.

5. *Simplify your equation.* For example, reduce the parentheses, combine like terms, and so forth.

6. Remember: *Whatever you do to a number, you can do to an unknown* (i.e., *x* or *y*).

7. *Trial and error may be okay.* Sometimes, you may have to plug in numbers to figure out how to solve an equation.

Many students have come to grips with algebra by treating it like a game and a positive mental challenge. Step by step, they follow a systematic procedure—and behold—they are led directly to a conclusion. The little *x*s and *y*s begin to make sense, and finding solutions becomes not only a challenge but also fun.

I would like to emphasize that each of the problem-solving suggestions offered in this chapter would work best for you if you practice them along with the relaxation techniques described earlier in this book, have a positive attitude toward math, and a willingness to try new patterns of thinking. Good luck!

Build Metacognitive Strategies

In this section, we explore your "cognitive" or thinking processes. Let's begin with a few simple questions: Have you ever observed how you think and how you make progress as you acquire new math concepts? If a concept is difficult to understand, do you consciously adjust your approach or adapt a different one to facilitate your comprehension and acquisition of this concept? Lastly, when you solve a problem, do you monitor which strategies are most effective, when they are appropriate to use, and how they should be applied so you can reach your goal?

Metacognition entails all this and more. It is your knowledge and in-depth understanding of how you think, learn, and solve problems, along with your ability to monitor, regulate, and evaluate the factors that influence them. You can think of metacognition as a management executive in charge of a large processing plant. If the executive manages his or her job well and continues to update, revise, evaluate, and oversee all the systems, then the plant operations will work smoothly and efficiently, and the plant will successfully accomplish its mission.

Individuals who are successful in math use metacognitive strategies all the time. These students constantly monitor and reflect on how they learn and progress in math. They check and recheck their work. They question why they got something wrong and how they can best improve their performance. Their metacognitive strategies help them to feel more confident about math and ensure their success.

Metacognitive strategies are best used before, during, and after you study and work out math problems. Self-questioning is an excellent way to improve your metacognition. Exercise 10-3 suggests questions for you to explore.

EXERCISE 10-3 Metacognitive Strategies for Success in Math

This exercise focuses on the three main stages of metacognition: the planning stage, the monitoring stage, and the reflection/evaluation stage. During each of these stages, there are important questions to ask yourself. Use the following questions to help you improve your metacognition skills:

The Planning Stage: Before you start a math problem-solving session, ask these questions:

1. What do I want to accomplish during this session?

2. Am I motivated to do this now?

3. How much time am I willing to devote to completing this work?

4. What prior information will help me?

5. On what should I work first?

6. What important procedures must I use to complete this task?

7. How should I divide up this task?

8. What sequence of steps should I follow?

9. How should I allocate my time with regard to the different things I must accomplish?

The Monitoring Stage: During your math problem-solving session, ask these questions:

1. Am I following my plan as I set it up before I started?

2. Is my plan working, or do I need to revise it and proceed differently?

3. Am I focused?

4. Is my motivation still up?

5. Am I still on track?

6. Do I understand the concepts and what I am doing?

7. Do I have to go back and reread my material?

8. Am I moving toward my goal for this session?

9. Have I missed something?

10. Do I have knowledge gaps that need to be filled in?

The Reflection/Evaluation Stage: After your math problem-solving session, ask these questions:

1. What worked well and what didn't work for me?

2. How can I check my work to make sure I really know and understand these concepts?

3. Did I check my work to make sure I did everything correctly?

4. Did my plan work out?

5. Is my plan worth using again?

6. What did I learn from doing this?

7. What should I do differently next time?

8. How can I apply the strategies I used in this session to others in the future?

9. What did I learn from this session that could help improve my thinking next time?

10. What do I have to review or study more to complete my understanding of these concepts?

Summary

You, too, can think like a mathematician. You needn't toil aimlessly when given problems to solve. In this chapter, I discussed the mathematician's logical problem-solving approach, some practical ways to integrate left- and right-brain thinking, and famous mathematician George Polya's classic problem-solving model. I also shared with you the secrets for dealing with word problems, a winning 12-step system for tackling word problems, and some tricks for solving algebra problems. I encouraged you to build your metacognitive abilities and I provided important questions to ask yourself during the planning, monitoring, and reflection/evaluation stages of metacognition. My goal for you is to meet the challenge of math—and to have fun doing it!

Math Is the Future

Do math . . . and you can do anything!

—National Council of Teachers of Mathematics Campaign

For many, success in math is synonymous with tackling fear. If you've come this far, you've proven your mettle. You've demonstrated toughness, a willingness to explore new directions, and a refusal to give up and accept the status quo. Such mental resilience is reminiscent of Psyche in Greek mythology. A mere mortal, Psyche faced fearsome tasks and challenges. In overcoming them, she was elevated to the status of a goddess.

In our everyday lives, the rewards for perseverance may not seem so exalted, but they are real nonetheless. Handling fear with dogged determination can pay off handsomely. The rewards abound. Math is the link.

No longer just the language of science, mathematics now contributes in direct and fundamental ways to business, finance, health, and defense. For students, it opens doors to careers. For citizens, it enables informed decisions. More than ever before, Americans need to think for a living; more than ever before, they need to think mathematically. —National Research Council, *Everybody Counts: A Report on the Future of Mathematics Education*

Math opens the doors to your future. To be an active, concerned member of this world, you must use the power of math. To be successful in school; to have a rewarding, stimulating career; to get the

" STICK TO ANIMALS, THOG. "

jobs you want; to be an involved citizen; to have a knowledge of personal finances, the nation's economy, and the technological advances of modern-day society—all these require you to have an understanding of math. This final section of our road map explores the importance of math in choosing your career, in assuring higher salaries, and in every aspect of today's ever-changing world.

Why Math?

If you used to think math was a bunch of numbers obscuring your appreciation of the world, you had it backward. Math is humankind's constant effort to describe (and put to use) the mysterious workings of the universe. Lacking systems of measurement, nonliterate cultures could not produce machines, manufactured goods, or reproducible structures. Eruptions and earthquakes, storms, and celestial events were attributed to superstitious, and therefore nonverifiable, causes.

How times change—and so, too, do the needs and demands of an increasingly complex society. Seismic analysis, plate tectonics, meteorology, and astrophysics have created order out of seeming chaos. But our incessant need for precise measurement and manipulation of data has outstripped anything ever before imagined.

Many students don't realize math is far more than the ability to calculate, memorize formulas, or solve equations. Math trains your mind to think logically and succinctly. It requires you to perceive patterns, observe relationships, clarify and critically analyze problems, deduce consequences, formulate alternatives, test conjectures, estimate results, and enhance your problem-solving abilities. By sharpening your reasoning and thinking skills, you can become more productive in every facet of your life.

Math provides you with the resources to comprehend the barrage of information communicated to you each day. It gives you the ability to be a critical reader of anything you read, from newspaper reports and research articles to insurance policies and loan documents. Math logic, reasoning, and thinking ability help you to ascertain possible risks or fallacies, to unearth biases, and to come up with suggestions and alternatives.

It's no wonder so many careers require math skills. Employers want to hire individuals who can solve problems, who can think clearly on the job, and who can deal with new ideas, ambiguity, and change. We are constantly being flooded with technological advances, new scientific discoveries, new knowledge. Are you prepared for this challenge?

The Mathematical Association of America (MAA) has an online essay series written by authors describing how their math background was useful in their careers. Dr. Richard Jarvinen, from Winona State University, wrote about the need for math when he did reliability studies and risk assessment for the NASA Space Shuttle program. Dr. Michael Monticino, from the University of Texas, used math in evaluating antisubmarine warfare tactics, and Dr. Michael J. Murray, a Java™ programmer who worked on the Olympic Internet Team, attributed his ability to query large sports statistics databases to his roots in math. Dr. Ellen Lentz was an animal science major and soon discovered much of modern agricultural development was attributable to experimental statistics. After years of using statistics to design experiments in veterinarian medicine, she now does the same work with human medicines. As a marine research associate at the University of Rhode Island, Allison DeLong used mathematics/statistics on projects related to fisheries stock assessment and the population dynamics of exploited fish stocks. For more exciting ideas on how useful math can be, check out MAA's Web site: www.maa.org/students/career.html.

Math-anxious students frequently ask, "Why do I have to take math?" or "Why do I have to learn algebra?" These questions came up so often I decided to interview students who enjoyed studying math to see how they would answer them. Here are some of their responses:

> "Name a good-paying job that doesn't require math."
>
> "Math affects everything I do: balancing my checkbook, checking over my bills, paying my credit cards, following my investments (returns, losses, etc.), shopping for bargains, buying a car, figuring out my taxes. It helps computing how much carpet to install in my bedroom, how many bundles of shingles my roof needs, how much crushed rock will cover my patio, in figuring out 30% off prices, totaling how many clock hours I put in at work this month, and if I want to cut a recipe in half, measuring ingredients."
>
> "Knowing math has saved me thousands when financing my car! I was able to figure out that those low-monthly payments would have cost me a fortune."
>
> "Besides the obvious reasons, such as figuring out my paycheck or comparison shopping, math has helped me to think in a logical manner. This, in turn, makes problem solving in my classes and in my personal life so much easier."
>
> "Math is one of the first courses where I was taught how to think and to reason. It's helped me think in a methodical way, step by step until I can draw a conclusion."
>
> "I figured out a long time ago that people with low math abilities are destined to be taken advantage of by insurance agents, car salespeople, investment advisers, and politicians. If you have reasonable math skills, you have a better chance to think for yourself and are less likely to be swindled. For example, recently I corrected a cashier who overcharged me by accidentally totaling my bill as $184 when it should have been $148."
>
> "One thing's for sure from a career standpoint. Job opportunities expand as your math proficiency increases. Besides the obvi-

ous Internet, computer, and electronic technology jobs, all supervisory, management, and administrative positions in the business world are requiring proficiency in math. Low-paying service sector jobs are the best options for those with little or no math skill."

"I know for a fact, doing math and math problems actually helps make me smarter. My thinking is so much clearer. I analyze situations I encounter in life better and I end up figuring out the best thing to do. In fact, my stock portfolio has doubled because of it!"

"Math describes what is going on around us. It answers why things do what they do."

"Algebra has helped me find out how much money I can put away today at the current interest rate to have a down payment on a new house in 5 years."

"Algebra has helped me in my nutrition major. For example, I needed it for an assignment where I was to provide nutritious meals for a family of four, using tables listing the protein, fat, cholesterol, carbohydrate, vitamin, and mineral content in quantities per gram, when the food I purchase is in pounds."

"I use algebra all the time. Just in the last few months, it has given me answers to: Should I buy whole life vs. term life insurance? Is a 4-year optional service contract on my new CD player really a good deal? How many cubic yards of fertilizer did I need for my vegetable garden? How much paint did I need to paint my living room and bedroom? Which interest loan on my car was the best deal?"

"Algebra forces you to think on your own, to become a critical thinker and a problem solver."

"Math is the language of choice for physics, chemistry, astronomy, geology, biology, psychological research and statistics, business and finance, electronics, computer programming, accounting, economics, and many other fields. Without it, you cannot begin to understand or describe these fields. With a basic understanding of this 'language,' you can begin to make sense of these subjects. Knowing just the basics opens many doors that would otherwise be closed."

"Algebra is the steppingstone for many higher-level science and math courses. It is analogous to learning to walk before running. Algebra gives us the beginning tools needed to solve many applied problems."

Today's Lucrative, Exciting Careers

William "Bill" Schrader, chairman of the board and CEO of PSINet, a commercial Internet service provider, knows how important math is in achieving success in today's technological world. In an interview, he was asked, "What are the most important aspects of mathematics needed in the workforce?" He responded, "It is really the ability to think and to imagine how to solve a problem . . . Everything comes down to mathematics . . . If employees turn out to be good problem solvers, then they progress rather rapidly—they can do anything, whether it's in sales, accounting, technology."

Whether you realize it or not, an understanding of mathematical problem-solving and reasoning ability is necessary to enter and advance in most jobs and careers. The National Research Council (NRC) reports 75% of all jobs require proficiency in basic algebra or geometry as a prerequisite for licensure or training. In addition, more than three-quarters of the nation's university degree programs require more advanced math such as calculus, discrete math, statistics, or comparable mathematics. An advanced level of math competence is one of the essential elements for comprehending the mathematical basis for the sciences, engineering, technology, computer programming, and business. Moreover, most, if not all, technical jobs require computer skills, and these in turn are largely dependent on math skills or mathematical thought processes. As with all graduate programs, math competence is a prerequisite for entering medical school. Beyond college algebra, trigonometry, and calculus, medical students must also have working knowledge of statistics, the major underpinning of all medical research. Students who avoid taking math are shut out of these lucrative career options. Please don't allow this to happen to you.

I often ask students who are overcoming math fears to complete this incomplete sentence: "If I were better in math _____." These are just a few of the responses I've received:

I'd be a Web designer.

I'd be an engineer.

I'd get a better job.

I'd be a financial planner.

I'd get my B.A. degree.

I'd be a doctor.

I'd finally pass the civil service exam.

I'd become a scientist.

I'd do better on the GRE and get into graduate school.

I'd pass the advancement exam for my job.

I'd feel great about myself and my future.

You needn't let math hold you back. I encourage you to continue to practice the strategies in this workbook and to follow your dreams. Don't close the doors on the exciting careers of today and the future. *You can succeed in math; you are a winner.*

Exercise 11-1 lists interesting and often high-paying careers requiring an understanding of mathematical reasoning and problem solving. See if you can answer some of the questions I've posed.

EXERCISE 11-1 Career Opportunities with Math

Here are more than 80 rewarding careers that utilize math skills and thinking ability. Try answering the following questions with the help of a career counselor or career resource materials. The *Occupational Outlook Handbook* from the U.S. Government Bureau of Labor Statistics is an excellent online career resource. It contains career descriptions, required education and training, and earning potentials. You'll find it at www.bls.gov/oco/. *The Riley Guide: Explore Career Options* is another helpful Web site to explore. Check it out at www.rileyguide.com.

What kind of work does each of these people do?

How is math used in this career?

What math courses would help me succeed in this career?

What salary would I expect to earn in this field?

accountant	aircraft design	astronaut
actuary	engineer	astronomer
aerospace engineer	airline navigator	attorney
agriculturalist	archeologist	audio engineer
agronomist	architect	automotive engineer

banker
biochemist
biomedical engineer
biometrician
broadcast and sound
    technician
business administrator
cancer research
    scientist
cartographer
chemical engineer
chemist
civil engineer
commercial pilot
company president/
    CEO
computer engineer
computer hardware
    engineer
computer music
    programmer
computer scientist
computer security
    specialist
computer support
    specialist
computer systems
    analyst
credit analyst
cryptographer
database
    administrator

design engineer
economist
electroneurodiagnostic
    technologist
electronic engineer
estimator
financial analyst
food scientist
geodetic surveyor
geologist
geophysical
    prospecting
    surveyor
geophysicist
health information
    technician
health technologist
help-desk technician
industrial engineer
industrial traffic
    manager
marine surveyor
marketing manager
mechanical engineer
meteorological
    technician
meteorologist
musicologist
network administrator
network systems
    analyst

nuclear medicine
    technologist
nuclear scientist
nurse
oceanographer
optometrist
orthopedic surgeon
photogrammetrist
physician
program analyst
psychologist
psychometrician
purchasing agent
radiologist
real estate broker
research assistant
school principal
software engineer
software quality
    assurance analyst
systems architect
systems developer
technical support
    specialist
technical writer
urban planner
weapon system analyst
Web designer/
    developer
Webmaster

Enjoy Greater Academic Success

The National Council of Teachers of Mathematics (NCTM) finds students who progress through mathematics are more likely to succeed in their education than those who avoid it. A study by the College Board shows math as an important monitor of academic success in college. Math is "the great equalizer" between high-income students and those who are disadvantaged, of a minority, or have low income.

The gap between these two groups in terms of their persistence and attendance in pursuit of a college degree essentially disappears when the latter group masters algebra and geometry in high school.

Score Higher on Exams

Many students who've overcome math anxiety find their math knowledge and positive attitude help on important tests. Craig, a psychology major, recently applied for a position with a large department store's online catalog division. He was required to take a preemployment assessment exam testing his math and problem-solving skills. He was nervous, as were the other applicants waiting for the exam to begin. While in the waiting room, Craig practiced the Deep Abdominal Breathing exercise described in Chapter 3 (and recorded on track three of the CD accompanying this book). During the assessment, he remained calm, self-assured, and confident. Not surprisingly, Craig did well on the test and got the job.

Poovi, a mid-level manager at a major industrial plant, returned to school part-time with the sole purpose of improving her math skills. Although she tried eight times, she was unable to pass her company's internal promotion exam, which stressed math and logical sequential patterning skills. After taking a few semesters of math, Poovi passed her company's test and was promoted. She is now the vice president of internal operations, earning a quite respectable six-figure salary.

The study of mathematics provides excellent preparation for advancement in many careers and jobs. According to the Institute of Education's research report entitled "Standardized Test Scores of College Graduates," mathematics majors have ranked the highest in test performance for the Law School Admissions Test (LSAT) and the Graduate Management Admissions Test (GMAT). High school students already know the important role of math for the Standard Achievement Test (SAT). But did you know training in mathematics also ensures you greater success on the Graduate Record Exam (GRE), an exam required by most universities for entrance into graduate programs?

Earn Higher Salaries

People who don't count won't count. —Anatole France

Because math is essential in engineering, business, science, economics, and all technological fields, individuals who enter these disciplines with a good grasp of mathematical concepts and reasoning ability have a greater likelihood of succeeding and progressing up the career ladder. There is a positive correlation between people's future income and both the number of math courses they take and how well they do in those courses.

Bachelor's degrees in engineering and computer science require several semesters of advanced mathematics beyond college algebra. Of all the degrees a student can receive at the undergraduate level, these are among the highest paying fields one can enter. Recent graduates in electrical and electronic engineering are receiving job offers with salaries starting at $55,000. Computer science graduates are in such high demand their starting salaries are even higher. According to Robert Half International, computer science job applicants may expect to be offered the following starting salaries:

database administrators	$70,000 to $88,000
network administrators	$43,000 to $62,500
network security administrators	$69,000 to $98,750
software engineers	$66,500 to $99,750
software installers/developers	$60,250 to $94,750
Webmasters	$60,250 to $81,000
Web developers	$54,750 to $81,500

Not bad for your first job out of college!

Math in the 21st Century

I was in college before I realized that mathematics is not cold. Mathematics is not like a piece of colorless cold marble—it has feeling, it has a warm temperature; mathematics has a beauty; it has a pulse; it has a heartbeat. —Bill Cosby, "Math: Who Needs It?" (PBS Video)

Not only is change an inescapable part of our lives, but the rate of change is accelerating at a prodigious pace. Change will continue unabated in the 21st century. A working knowledge of math is an indispensable survival skill. In this section, we'll explore major areas in which an understanding of mathematical concepts is essential. Can you add more?

Math in Science, Industry, and Technology

1. Math is the language of all technological, scientific, and industrial research and investigation.
2. Calculating escape velocities, trajectories, and orbital data in the space program.
3. Analyzing bandwidth capacity exchange for the telecommunications industry.
4. Detecting high-tech crime in data retention and preservation, in threat assessment and prevention, and in protecting electronic commerce.
5. Developing encrypted communication devices.
6. Developing high-performance fiber-optics illumination devices.
7. Designing automated systems and robotics.
8. Locating natural gas reserves and mineral deposits.
9.
10.

Math in Transportation, Weather, and Disaster Relief

The Universe is a grand book which cannot be read until one first learns to comprehend the language and becomes familiar with characters of which it is composed. It is written in the language of mathematics. —Galileo

1. Forecasting weather changes, such as tornado and hurricane warnings.
2. Analyzing building stress points during earthquakes to allow for reinforcement protection.

3. Analyzing floodplains and damage control.
4. Analyzing weather data transmissions from satellites orbiting the earth.
5. Designing airliners, trains, and cruise ships.
6. Analyzing traffic patterns to plan for road construction.
7. Planning airline flight schedules and flight paths.
8. Figuring out your gasoline needs for long automobile trips.
9. Building dams, tunnels, and bridges.
10.
11.

Math in Health Issues

1. In the Human Genome Project, assembling sequenced fragments of the human genome using computational biology and sequence analysis software.
2. Researching viral load assay procedures for measuring AIDS patients' virus levels to pinpoint when drug resistance begins.
3. Designing and updating laparoscopes, endoscopes, and computer-assisted scanning devices such as the VQ (Ventilation/Perfusion) for identifying pulmonary emboli and others such as the CT, the PET, and the MRI.
4. Studying retroviruses and monoclonal antibodies.
5. Understanding the health hazards involved in acid rain, the greenhouse effect, and waste management.
6. Determining the nutritional value of foods.
7. Evaluating outrageous scientific claims and discriminating among rational and reckless claims.
8. Accurately measuring drug dosages; errors can be fatal.
9. Understanding research studies on placebo effects of drugs.
10.
11.

Math in Home Entertainment Systems

1. Developing high-definition television (HDTV), interactive television, and DVDs.
2. Making advances in broadband technology for high-speed access to home users.

3. Developing the next-generation receivers and decoders for digital satellite, cable, and terrestrial television.
4. Developing interactive multimedia software.
5. Designing interactive game software with its game worlds and architecture along with computer modeling and animation.
6.
7.

Math in Music, Art, and Design

Music is the pleasure the human mind experiences from counting without being aware that it is counting. —Gottfried Leibniz

Architecture is akin to music in that both should be based on the symmetry of mathematics. —Frank Lloyd Wright

1. Counting is basic to all musical performance.
2. Producing notes on a musical scale by taking the integral ratios of a string's length.
3. Determining the difference between a whole note, a quarter note, or a sixteenth note involves conceptualizing ratios and proportions.
4. Understanding fractions and multiples is needed for playing different rhythms such as three-quarter time versus four-quarter time.
5. Using mathematical concepts to create artistic patterns and designs. Famous artists such as Albrecht Dürer, Leonardo da Vinci, and M. C. Escher relied on an understanding of math to create their masterpieces.
6. In the ancient Japanese art of origami, using mathematical ideas of lines and points of symmetry, congruences, ratios, and proportions of shapes.
7. Applying mathematical concepts and measurements in architectural design.
8. In the design of Native American Indian rugs, using lines of symmetry, reflection, congruences, and accurate geometric proportions.

9. Architects and interior designers using computer-aided drafting/designing for three-dimensional modeling and solid modeling applications.
10.
11.

Math in Civic Issues

1. Evaluating national political polls and presidential election data returns.
2. Understanding changes in the federal and state tax codes.
3. Understanding the U.S. Census Bureau reports and data on population growth.
4. Evaluating the nation's nuclear armament and defense systems.
5. Evaluating financial news data on the gross national product, the consumer price index, the nation's money supply, unemployment projections, and new home construction costs.
6. Determining electrical, water, and gas consumption needs of your city, state, or the nation.
7.
8.

Math in Leisure Time Activities

1. Exchanging U.S. dollars to foreign currencies when traveling.
2. Figuring sports data and scoring.
3. Playing games of strategy, chance, and probability.
4. Figuring out puzzles.
5.
6.

Math in Finances

1. Setting a household budget and computing taxes.
2. Developing investment strategies.
3. Investing in the stock market and understanding mutual fund and stock annual reports and prospectuses.

4. Updating a stock portfolio and charting investment yields.
5. Understanding the effects of inflation on income and investments.
6. Comparing insurance policies; calculating insurance payments.
7. Figuring out mortgage rates and payments.
8. Computing payroll taxes and deductions.
9. Analyzing profit and loss statements.
10. Comparing the rates of mortgage, home improvement, or car loans.
11.
12.

Go for It!

What you have to make them see—more than anything—is the future—which is mathematics. —Bill Cosby, "Math: Who Needs It?" (PBS Video)

It's up to you to take action and jump into the wonderful world of math. You have the tools to succeed. Take it one positive step at a time, assertively move forward, and you will accomplish your goals.

EXERCISE 11-2 Collaborative Learning: Make a Commitment to Take Action

List five action steps you will take in the next 6 months as a result of completing this workbook. Make sure to write down your target date for each step taken. Share your action steps with a small group of classmates.

Action Steps   **Target Date**

1.

2.

3.

4.

5.

Summary

Math is the one constant in all sciences' efforts to understand, manipulate, and control natural phenomena. But beyond its necessity in science, mathematics is essential in all aspects of our lives. This chapter explored the importance of math in training us to think clearly. It answered the question "Why study math?" It also showed us that math can open doors to lucrative career opportunities, greater academic success, and higher scores on entrance and job placement exams. We also examined the applications of math in science, technology, transportation, weather, health issues, home entertainment and computer systems, civic issues, finance, leisure time, music, art, and culture.

Congratulations! You have arrived. You have completed the major steppingstones along the road to math achievement. You've made a commitment to succeed. You are a winner!

REFERENCES

Ashcraft, M. H. (1995). Cognitive psychology and simple arithmetic: A review and summary of new directions. *Mathematical Cognition, 1*, 3–34.

Ashcraft, M. H., & Kirk, E. P. (2001). The relationship among working memory, math anxiety, and performance. *Journal of Experimental Psychology: General, 130*, 224–237.

Ashcraft, M. H., & Krause, J. A. (2007). Working memory, math performance, and math anxiety. *Psychonomic Bulletin & Review, 14*(2), 243–248.

Bandura, A. (1997). *Self-efficacy: The exercise of control.* New York: Freeman.

Beilock, S. L., & Carr, T. H. (2005). When high-powered people fail: Working memory and "choking under pressure" in math. *Psychological Sciences, 16*(2), 101–105.

Brannon, L. (2008). *Gender—Psychological perspectives* (5th ed.). Boston: Allyn & Bacon.

Cates, G., & Rhymer, K. N. (2003). Examining the relationship between mathematics anxiety and mathematics performance: An instructional hierarchy perspective. *Journal of Behavioral Education, 12*(1), 23–34.

Coué, E. (1922). *Self-mastery through conscious auto-suggestion.* London: Allen & Unwin.

Cowan, N. (2008). Working memory. In N. J. Salkind (Ed.), *Encyclopedia of educational psychology* (pp. 1015–1016). Thousand Oaks, CA: Sage Publications.

Dar-Nimrod, I., & Heine, S. J. (2006). Exposure to scientific theories affects women's math performance. *Sciences, 314*(5798), 435.

Drake, R. A., & Seligman, M. E. P. (1989). Self-serving biases in causal attribution as a function of altered activation asymmetry. *International Journal of Neuroscience, 45*, 199–204.

Drake, R. A., & Ulrich, G. (1992). Line bisecting as a predictor of personal optimism and desirability of risky behavior. *Acta Psychologica, 79*(3), 219–226.

Dunn R., & Dunn, K. (1978). *Teaching students through their individual learning styles: A practical approach.* Englewood Cliffs, NJ: Prentice Hall.

Dunn, R., & Dunn, K. (1979). *Student learning styles: Diagnosing and prescribing programs.* Reston, VA: National Association of Secondary School Principals.

Eccles, J. S. (1989). Bringing young women to math and science. In M. Crawford & M. Gentry (Eds.), *Gender and thought: Psychological perspectives* (pp. 36–58). New York: Springer-Verlag.

Eccles, J. S., & Jacobs, J. E. (1986). Social forces shape math attitudes and performance. *Signs, 11,* 367–380.

Ellis, A. (1998). *Guide to rational living* (3rd Rev. ed.). North Hollywood, CA: Wilshire Books.

Feingold, A. (1988). Cognitive gender differences are disappearing. *American Psychologist, 43,* 95–103.

Feingold, A. (1994). Gender differences in variability in intellectual abilities: A cross-cultural perspective. *Sex Roles, 30,* 81–92.

Fennema, E. (1980). Sex-related difference in mathematics achievement: Where and why. In L. H. Fox et al. (Eds.), *Women and the mathematical mystique* (pp. 76–93). Baltimore: Johns Hopkins University Press.

Hanna, G. (1989). Mathematics achievement of girls and boys in eighth grade: Results from twenty countries. *Educational Studies in Mathematics, 20,* 225–232.

Haring, N. G., & Eaton, M. D. (1978). Systematic instructional technology: An instructional hierarchy. In N. G. Haring, T. C. Lovitt, M. D. Eaton, & C. L. Hansen (Eds.), *The fourth R: Research in the classroom* (pp. 23–40). Columbus, OH: Merrill.

Helping students develop math power. (1991, April 18). *Guidepost,* p. 11.

Huang, J. (1993). An investigation of gender differences in cognitive abilities among Chinese high school students. *Personality and Individual Differences, 15*(6), 717–719.

Hyde, J. S., Fennema, E., & Lamon, S. J. (1990). Gender differences in mathematics performance: A meta-analysis. *Psychological Bulletin, 107,* 139–155.

Hyde, J. S., Lindberg, S. M., Linn, M. C., Ellis, A. B., & Williams, C. C. (2008). Gender similarities characterize math performance. *Sciences, 321*(5888), 494–495.

Jackson, C. D., & Leffingwell, R. J. (1999). The role of instructions in creating math anxiety in students from kindergarten through college. *The Mathematics Teacher, 92*(7), 583.

Kiefer, A. K., & Sekaquaptewa, D. (2007). Implicit stereotypes, gender identification, and math-related outcomes: A prospective study of female college students. *Psychological Science, 18*(1), 13–18.

Kimball, M. (1995). *Feminist visions of gender similarities and differences.* New York: Haworth Press.

Lummis, M., & Stevenson, H. W. (1990). Gender differences in beliefs and achievement: A cross-cultural study. *Developmental Psychology, 26*(2), 254–263.

Magnesen, V. (1983). A review of findings from learning and memory retention studies. *Innovation Abstracts, 5* (No. 25).

"Math: Who Needs It?" (1991). PBS Video.

Michaelson, M. T. (2007). An overview of dyscalculia. *Australian Mathematics Teacher, 63*(3), 17–22.

National Center for Learning Disabilities, http://www.ncld.org

National Research Council. (1989). *Everybody counts, a report to the nation on the future of mathematics education.* Washington, DC: National Academy Press.

No gender differences in math performance. (2008, July 24). *Science-Daily.* Retrieved from http://www.sciencedaily.com/release/2008/07/080724192258.htm

Pauk, W., & Owens, R. J. Q. (2008). *How to study in college.* (9th ed.). Boston: Houghton Mifflin.

Polya, G. (1985). *How to solve it—A new aspect of mathematical method* (2nd ed.). Princeton, NJ: Princeton University Press.

Richardson, A. (1969). *Mental imagery.* New York: Springer.

Richardson, F. C., & Suinn, R. M. (1972). The mathematics rating scale: Psychometric data. *Journal of Counseling Psychology, 19*(6), 551–554.

Robbins, A. (1997). *Unlimited power: The new science of personal achievement.* New York: Simon & Schuster.

Saults, S. J., & Cowan, N. (2007). A central capacity limit to the simultaneous storage of visual and auditory arrays in working memory. *Journal of Experimental Psychology: General, 136*(4), 663–684.

Schaum's Outline Series. New York: McGraw-Hill.

Sells, L. (1978). Mathematics—A critical filter. *The Science Teacher, 45*(2), 28–29.

Sells, L. (1980). The mathematics filter and the education of women and minorities. In L. H. Fox et al. (Eds.), *Women and the mathematical mystique* (pp. 66–75). Baltimore: Johns Hopkins University Press.

Sharma, M. (2003). Dyscalculia. *BBC Skillswise expert column.* Retrieved from http://www.bbc.co.uk/skillswise/tutors/expertcolumn/dyscalculia/

Skaalvik, E. M., & Rankin, R. J. (1994). Gender differences in mathematics and verbal achievement, self-perception and motivation. *British Journal of Educational Psychology, 64*(3), 419–428.

The resource manual for counselors/math instructors, math anxiety, math avoidance, reentry mathematics. (1980). Prepared for the U.S. Dept. of Education, the Fund for the Improvement of Postsecondary Education.

Project Director: Sheila Tobias. Washington, DC: The Institute for the Study of Anxiety in Learning, The Washington School of Psychiatry.

Trujillo, K. M., & Hadfield, O. D. (1999). Tracing the roots of mathematics anxiety through in-depth interviews with preservice elementary teachers. *College Student Journal, 33*(2), 11.

Vaidya, S. R. (2004). Understanding dyscalculia for teaching. *Education, 124*(4), 717–720.

Wurtman, J. J. (1988). *Managing your mind and mood through food.* New York: HarperCollins.

Yee, D. K., & Eccles, J. S. (1988). Parent perceptions and attributions for children's math achievement. *Sex Roles, 19*, 317–333.

Bluman, A. G. (2005). *Math word problems demystified*. New York: McGraw-Hill.

Davis, M., Eshelman, E. R., McKay, M., & Fanning, P. (2008). *The relaxation and stress reduction workbook*. Oakland, CA: New Harbinger.

Fanning, P. (1994). *Visualization for change*. Oakland, CA: New Harbinger.

Jacobs, H. (1994). *Mathematics: A human endeavor: A book for those who think they don't like the subject*. San Francisco: Freeman.

LearningExpress Editors. (2004). *1001 math problems (skill builders practice)*. New York: LearningExpress.

LearningExpress Editors. (2008). *Algebra success in 20 minutes a day* (3rd ed.). New York: LearningExpress.

Pappas, T. (1991). *More joy of mathematics—Discovering mathematics all around you*. San Carlos, CA: Wide World Publishing Tetra.

Paulos, J. A. (2001). *Innumeracy—Mathematical illiteracy and its consequences*. New York: Hill & Wang.

Polya, G. (1985). *How to solve it—A new aspect of mathematical method* (2nd ed.). Princeton, NJ: Princeton University Press.

Singh, S. (1997). *Fermat's enigma—The epic quest to solve the world's greatest mathematical problem*. New York: Walker.

Smith, R. M. (1999). *Mastering mathematics: How to be a great math student*. Pacific Grove, CA: Brooks/Cole.

Walter, T., Siebert, A., & Smith, L. (2000). *Student success: How to succeed in college and still have time for your friends* (8th ed.). Troy, MO: Harcourt.